MODERN PRACTICE OF
LIQUID CHROMATOGRAPHY

CONTRIBUTORS

KARL J. BOMBAUGH, Waters Associates, Inc., Framingham, Massachusetts

S. H. BYRNE, JR., Instrument Products Division, E. I. du Pont de Nemours & Company, Wilmington, Delaware

DENNIS R. GERE, Varian Aerograph, Walnut Creek, California

ISTVÁN HALÁSZ, Institut für Physikalische Chemie der Universtität, Frankfurt am Main, Germany

RICHARD A. HENRY, Instrument Products Division, E. I. du Pont de Nemours & Company, Wilmington, Delaware

BARRY L. KARGER, Department of Chemistry, Northeastern University, Boston, Massachusetts

JOSEPH J. KIRKLAND, Industrial and Biochemicals Department, E. I. du Pont de Nemours & Company, Wilmington, Delaware

JOHN A. SCHMIT, Instrument Products Division, E. I. du Pont de Nemours & Company, Wilmington, Delaware

CHARLES D. SCOTT, Oak Ridge National Laboratory, Oak Ridge, Tennessee

LLOYD R. SNYDER, Union Oil Research Center, Union Oil Company of California, Brea, California

MODERN PRACTICE OF
LIQUID CHROMATOGRAPHY

Edited by

J. J. KIRKLAND

Industrial and Biochemicals Department

E. I. du Pont de Nemours & Company

Wilmington, Delaware

Wiley-Interscience, a Division of John Wiley & Sons, Inc.

New York · London · Sydney · Toronto

This book is dedicated to a good friend who introduced me to the delights of gas chromatography in 1954, the late Steven Dal Nogare, a vigorous proponent of chromatography and one of the founders of the Chromatography Forum.

J.J.K.

PREFACE

The oldest of the chromatographic methods, column liquid chromatography, currently is experiencing a renaissance which is re-establishing the reputation of this technique as one of the most useful available to the scientist. Although predating gas chromatography by almost a half century, liquid chromatography has often been relegated to a "last ditch" status because of the inconvenience and time-consuming techniques traditionally associated with this approach. Within the last four years, however, there has been a sharp revival of interest, resulting primarily from advances which have greatly speeded up separations and improved the efficiency of the method. These gains have been due to the development of apparatus for high-speed, high-performance work and to improved column design based on theoretical insights gleaned mainly from gas chromatographic studies. Refinement of the technique to the point where the time and efficiency of separations rival those of gas chromatographic analyses ensures that liquid chromatography will now enjoy widespread application as an analytical separations tool.

The lack of coordinated information on the new techniques of high-performance, high-speed liquid chromatography prompted The Chromatography Forum of the Delaware Valley and the Instrument Products Division of E. I. du Pont de Nemours and Co. to cosponsor a comprehensive course entitled the "Modern Practice of Liquid Chromatography" in Wilmington, Delaware, on April 6–8, 1970. This book, an expanded version of the course material, represents the first effort to bring together existing information on the newer techniques of liquid chromatography. Each of the outstanding teachers and research workers contributing chapters has presented the most important and most recent developments in his own field, using appropriate references to cover other items. Thus, the book is not a collection of research or review articles, but contains critical accounts of the various subjects.

I have attempted to foster presentations which emphasize the teaching aspects. It is hoped that the novice can use this book as a starting point to begin his studies in high-speed liquid chromatography. The critical insights and up-to-date information should also be of value to those already working in the field.

Written by almost a dozen experts, the book is developed in a manner that stresses continuity. The format has been standardized to make it easier for the reader to follow and relate the various topics. For instance, the terminology and symbols used are common throughout the book.

Although an attempt has been made to minimize overlapping among the various contributors, certain subjects have been treated by more than one author with different conclusions. These varying opinions have been allowed to remain, since they reflect the fact that certain areas still need further development before a single approach can be accepted widely.

The book is divided into three main sections. Part I describes the latest concepts in liquid chromatographic theory, current equipment, including detectors, and the unique features of the carrier in liquid chromatography. Part II contains chapters describing practice and techniques in the four areas of column chromatography: liquid-liquid, liquid-solid, gel permeation, and ion exchange. Included in this section is an interesting and thought-provoking overview of certain aspects of liquid chromatography. Part III contains chapters describing various applications of high-performance liquid chromatography. The chapters in this section are written by representatives of instrument company application laboratories. These workers are currently developing the widest range of applications of any in the field. Some of these authors of Part III have concentrated their efforts in particular areas; consequently, their chapters necessarily reflect their own work. Although illustrations in all areas are not included, an attempt has been made to give insight into the practical approaches to various types of separation problems of current interest. Teaching aspects have again been emphasized in this part.

It is not possible to publish a book of this type without the generous cooperation of others. I am greatly indebted to Miss Mildred Syvertsen for her tireless and competent assistance in the preparation of the manuscript. Sincere thanks go to several members of the Chromatography Forum Executive Committee, to some of my laboratory colleagues, and to others who have given generously of their time to review the chapters in the book. Invaluable help was furnished by Gerald R. Umbreit, Aaron E. Wasserman, Rex E. Nygren, Daniel O'Leary, Lyle H. Phifer, Joseph J. DeStefano, Richard L. Fisher, Robert E. Leitch, John H. Knox, Lloyd R. Snyder, Karl J. Bombaugh,

Donald D. Bly, and Harold L. Suchan. Particular recognition and thanks are due also to Herman R. Felton, who was chairman of the committee organizing the course on liquid chromatography. Without his direction and energy this book would not have been possible.

Although producing this book has represented considerable work, I have thoroughly enjoyed the opportunity of closely associating with the various authors. They have been most cooperative and generous in their efforts to make the book a success. Probably all of the authors would agree that the book should be dedicated to our wives, who have been infinitely patient with us.

Joseph J. Kirkland

Wilmington, Delaware

November, 1970

CONTENTS

PART ONE FUNDAMENTALS OF HIGH-SPEED LIQUID CHROMATOGRAPHY

PART TWO   THE PRACTICE OF LIQUID CHROMATOGRAPHY

MODERN PRACTICE OF
LIQUID CHROMATOGRAPHY

PART ONE
FUNDAMENTALS OF HIGH-SPEED LIQUID CHROMATOGRAPHY

CHAPTER 1

The Relationship of Theory to Practice in High-Speed Liquid Chromatography

Barry L. Karger

A. INTRODUCTION

Over the last few years there has been a rebirth of interest in liquid chromatography with the achievement of rapid separations previously thought to be unattainable. This development has been due in large measure to the improvement in column efficiency and the use of high mobile-phase velocities. In this chapter these new developments will be described and their relationship to the fundamentals of chromatography will be shown.

The chapter will be divided into several sections. First some general aspects of chromatography will be briefly discussed. Then the fundamentals of chromatography, in particular the principles of the achievement of separation, will be examined. This discussion will require an examination of selectivity and efficiency in chromatography. Then optimization procedures in high-speed liquid chromatography will be discussed. Finally, several areas in need of development in this field will be indicated.

It is worthwhile to discuss briefly the role of theory in chromatography. Although the chromatographic process is exceedingly complex, it is possible to devise simple models and equations which closely approximate chromatographic separation. An understanding of these

Acknowledgment. The author gratefully acknowledges National Institutes of Health support of liquid chromatographic research at Northeastern, on which this article has been based in part. Also, some portions of this chapter are due to discussions with Professor Halász and Dr. Engelhardt; the author is most grateful for their stimulation.

simple equations can be of importance to the practitioner. For example, if insufficient resolution is obtained
in a given chromatogram, a worker needs to select a new
set of conditions to improve the separation. He can
select from literally an infinite number of variations,
and given sufficient time he should be able to pick a
reasonable set of conditions to obtain the best separation (adequate resolution in minimum time). However,
with a little knowledge of the fundamentals of separation in chromatography, he can logically select the new
set of conditions and decrease significantly the amount
of time and effort required to find the best separation
conditions. Indeed, his selection of the initial set of
conditions can be logically controlled by his understanding of simple theory. In addition, a worker with
some knowledge of the fundamentals can evaluate new
chromatographic techniques as they are developed and
reported. In this chapter we will present these simple
principles and show how they may be applied to real
separation situations.

Classical column liquid chromatography is a well-
established separation method (1). A column with a
diameter of roughly 1/2 in. is filled with packing
material such as alumina having a particle diameter of
100-150 μ. The mobile phase slowly flows through the
column by means of gravity. Samples are collected in
fractions and subsequently analyzed. Low column efficiencies and long separation times are characteristic
of this traditional method.

In the new method of operation small-diameter columns
(ca. 1-3 mm) and solid supports of particles with diameters less than 50 μ are used. At times specially
designed particles (porous layer beads) have been successfully employed. The eluent is pumped through the
column at a high flow rate (ca. 1-5 ml/min), resulting
in a significant pressure drop (ca. 1000 psi for a 1-
meter column). Figure 1.1 shows a three-component
separation obtained in less than 1 min with high-speed
liquid chromatography (LC), using a solvent velocity of

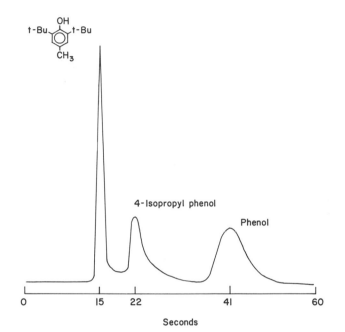

Figure 1.1. Three-component phenol mixture separation (3 μl injection) on 0.2% β,β'-oxydipropionitrile coated on 30 μ surface-etched beads (Corning Glass Works Co.). Column length = 50 cm, column diameter = 3 mm, mobile phase = n-heptane, velocity = 3.2 cm/sec (2).

3.2 cm/sec. The separation is a simple one, but it illustrates the speed possible in LC.

Gas chromatography (GC) is a very powerful separation method for volatile samples. However, components with molecular weights above roughly 300, thermally degradable species, and ionic species cannot be eluted. Liquid chromatography finds its greatest application in the separation of these types of samples. More specifically, components with molecular weights in the range of 300 to 1000-2000 (omitting gel permeation), as well

as ionic (ion-exchange) and thermally degradable species
(column temperature ~25°C), can be successfully chromatographed. Thus GC and high-speed LC complement one
another.

Many people who enter the field of high-speed LC have
had some experience with GC. Although this experience
can be useful, a word of caution is necessary. There
are some fundamental differences between the two techniques, and a person who is not aware of these differences may encounter problems in a first venture into
high-speed LC. For example, in GC extra-column effects,
such as slow injection and dead volume, are often neglected; however, in LC, especially with high-efficiency
columns, it is not possible to ignore these effects. In
this chapter some of these fundamental differences between gas and liquid chromatography will be pointed out.

B. GENERAL CHROMATOGRAPHY

1. Definitions and Descriptions

Chromatography basically involves separation due to
differences in the equilibrium distribution of sample
components between two immiscible phases. One of these
phases is a moving or mobile phase, and the other is a
stationary phase. The sample components migrate through
the chromatographic system only when they are in the
mobile phase. The velocity of migration is a function
of equilibrium distribution, with the components having
distributions favoring the stationary phase migrating
slower than those having distributions favoring the
mobile phase. Separation then results from different
velocities of migration as a consequence of differences
in equilibrium distributions.

There are three modes of chromatographic operation
-- elution, frontal, and displacement. In high-speed
liquid chromatography at the present time only the elution mode has been used. The frontal and displacement
modes are less often used than elution in all types of
chromatography and therefore will not be discussed here.
In elution the sample components are placed (injected)

at the beginning of the chromatographic system (bed) and development occurs through the bed. If the system is composed of a column, the components elute from the column according to their distributions between the stationary and the mobile phases, the concentration distribution being most often symmetrical and Gaussian. Symmetrical peaks occur when very small sample sizes are used (so that the distribution isotherm is linear). The greatest efficiency occurs in this case, and consequently in high-speed LC sample sizes are maintained very small.

Chromatographic methods can be classified according to the type of mobile and stationary phases selected. Gas chromatography encompasses those methods in which the mobile phase is a gas; liquid chromatography, those in which the mobile phase is a liquid. The different stationary phases give rise to the names liquid-solid chromatography (LSC) and liquid-liquid chromatography (LLC). Liquid chromatographic methods can also be classified according to the mechanism of retention---ion-exchange chromatography, adsorption chromatography, partition chromatography, and exclusion chromatography. All these mechanisms are well known with one exception: the exclusion mechanism, which forms the basis of gel permeation (GPC) or gel filtration chromatography. In this method the stationary phase is a porous solid, the size of the pores being similar to the dimensions of the component molecules. Separation results because of differences in the sizes of the sample molecules; those that are small enough are able to penetrate the porous matrix, whereas the larger components remain in the interstitial regions between the particles. Consequently, the largest components elute first, followed by those of smaller sizes. The method is used chiefly in the fractionation of polymers and proteins. Chapter 7 in this book will discuss GPC in detail.

Finally, liquid chromatography has been performed in a column and on an open bed (paper chromatography and thin-layer chromatography, TLC). High-speed LC has been performed almost totally in columns; however, thin-film chromatography was introduced recently as a high-speed form of thin-layer chromatography (<u>3</u>).

2. <u>Retention</u>

As we have already noted, if sample sizes are sufficiently small, symmetrical Gaussian peaks are obtained
as the components elute from the column. Under such
conditions it is possible to relate directly the time of
elution of the peak maximum to the equilibrium distribution coefficient. This time, called the retention time,
t_R, is illustrated in Figure 1.2. The assumption is
made that the average solute molecule maintains distribution equilibrium in its travel down the chromatographic

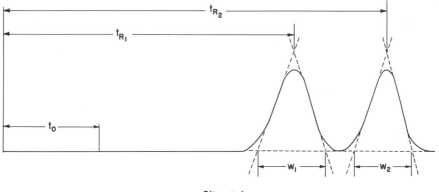

$$R_s = \frac{2(t_{R_2} - t_{R_1})}{w_1 + w_2}$$

<u>Figure 1.2.</u> Chromatogram illustrating the defini-
tions of retention time, t_R; nonsorbed time, t_0;
band width w; and resolution, R_s.

column. When a symmetrical peak is obtained, the retention time is independent of the amount of sample injected. Under these conditions the distribution isotherm
is linear (i.e., an increase in sample amount in the
mobile phase results in a proportionate increase of
sample amount in the stationary phase), the method being
called linear elution chromatography. As already noted,

these are the conditions that we strive to achieve in high-speed liquid chromatography.

Since the elution time will be a function of the velocity of the mobile phase, a more fundamental parameter of retention is the retention volume, V_R. The retention volume of a sample component represents the number of milliliters of mobile phase that must be passed through the column to elute the sample. The retention volume is simply equal to the mobile-phase flow rate, F, times t_R. The flow rate is, of course, directly proportional to the mobile-phase velocity, v, the proportionality constant being the cross-sectional area of the column. In gas chromatography flow rate and velocity are functions of the position in the column, since gases are compressible. At the pressures used in high-speed LC, on the other hand, the liquids are incompressible and therefore the velocity and the flow rate are constant in the column. In other words, no correction need be made for compressibility in LC. It is difficult to generalize the pressure at which a compressibility correction would be necessary; a good approximation, however, is between 8000 and 10,000 psi. This pressure is well above that ordinarily used in high-speed LC (ca. 1000-3000 psi).

Figure 1.2 also shows the time for elution of a non-retained component,* t_O; $t_O F$ gives the elution volume for a nonretained component, V_M which is the total volume of the mobile phase in the column (dead volume) if extra-column volumes are negligible (GPC is not included here). The dead volume is simply a constructional aspect of the column and does not directly contribute to the separation. Dead volume is considered in detail in Section C when we deal with the optimization of separation.

---

*There are a number of ways of determining t_O. The simplest method is to inject a compound closely related to the mobile phase (e.g., pentane when hexane is the mobile phase), detecting with a differential refractometer.

   The fundamental retention equation for any chromatographic process is given in equation 1.1:

$$V_R = V_M + KV_S \qquad (1.1)$$

where K is the equilibrium distribution coefficient (concentration of sample in stationary phase/concentration
of sample in mobile phase), and V_S is the volume of stationary phase (liquid volume, surface area, or weight of
absorbent). Often retention is given in terms of V_N,
the net retention volume,

$$V_N = V_R - V_M = KV_S \qquad (1.2)$$

   An important retention parameter in chromatography is
the capacity factor, k', defined as

$$k' = K \frac{V_S}{V_M} = \frac{\text{amount of solute in stationary phase}}{\text{amount of solute in mobile phase}} \qquad (1.3)$$

The capacity factor, which is basically the ratio of the
amounts of the sample component in the two phases, measures the extent of chromatographic distribution behavior
that the sample undergoes in its travel through the column. Combining equations 1.1 and 1.3 gives

$$V_R = V_M(1 + k') \qquad (1.4)$$

Here we see that, when k' = 0, the retention volume is
equal to V_M.

   Since retention time can be read directly on the chromatogram, we shall return to using this parameter,
assuming that the flow rate is constant. Equation 1.4
can be converted to retention and, when rearranged, gives

$$k' = \frac{t_R - t_o}{t_o} \qquad (1.5)$$

The capacity factor is simply the net retention time
relative to the nonsorbed time and can be determined
directly from the chromatogram.

   The time of a nonsorbed component, t_o, is equal to
the column length divided by the mobile-phase velocity,
L/v. Therefore equation 1.5 can be changed to

$$t_R = \frac{L}{v} (1 + k')$$ (1.6)

This equation gives the fundamental relationship between retention and equilibrium, column length, and mobile-phase velocity. Doubling the length of the column doubles the retention time, whereas doubling the velocity of the mobile phase halves the retention time.* The retention time depends also, through the capacity factor, on the relative amounts of stationary and mobile phase. Thus, in LLC a larger amount of stationary liquid phase results in a longer retention time. The final factor controlling retention is the equilibrium distribution, K. The larger this value, the longer is the retention time. It is worthwhile keeping these parameters in mind in our later discussions, for analysis time must often be sacrificed in the interests of efficiency and adequate separation.

C.   GENERAL SEPARATION PRINCIPLES

1.   Resolution

An understanding of separation in chromatography requires first a quantification of the degree of separation, and this can be done through the resolution, R_s. If a two-component chromatogram is examined, it is recognized immediately that two characteristics determine the degree of band overlap--the distance between the peak maxima and the widths of the bands. Figure 1.3 shows the influence of these two characteristics on resolution. In the top chromatogram, the bands overlap.

---

*For a given flow rate the velocity of the mobile phase will be dependent on the diameter of the column. Since the columns are cylindrical, the ratio of velocities for two columns will be a function of the squares of their radii. Thus, changing from a 3-mm- to a 1-mm-diameter column increases the velocity by a factor of 9 for a constant flow rate.

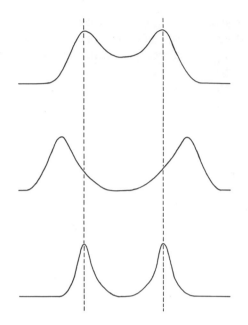

Figure 1.3. Chromatograms illustrating the influence of peak-to-peak separation and band widths on
resolution.

In the middle chromatogram, the distance between the
peak maxima has been increased while the band widths are
maintained constant. In the lower chromatogram, the
peak-to-peak distance is maintained constant but the
band widths are narrowed. It is clear that the peak
maxima should be well separated and the bands should be
narrow.

A reasonable measure of resolution is the following:

$$R_s = 2 \left(\frac{t_{R_1} - t_{R_2}}{w_1 + w_2} \right) \qquad (1.7)$$

where w_1 and w_2 are the band widths determined by the
intersection of the tangents of the inflection points of
the Gaussian peaks with the base line. The calculation
of the resolution is shown diagrammatically in Figure
1.2. It is worth pointing out that equation 1.7 is
simply a means of measuring the degree of separation of

two-component systems. The larger the value of the resolution the better the bands are separated. No assumptions enter into equation 1.7 except that the peaks are symmetrical (linear elution).*

The peak width for Gaussian bands is equal to 4σ, with σ the standard deviation of the Gaussian. Equation 1.7 then becomes, assuming equal peak widths for both components,

$$R_S = \frac{t_{R2} - t_{R1}}{4\sigma} \qquad (1.8)$$

When R_S equals 1, the peak-to-peak separation is equal to 4σ (this is called a 4σ separation). Since 2σ band widths will be involved for each peak, there will be only 2% band overlap. For high-speed separations $R_S = 1$ is taken as a satisfactory separation; a 6σ separation ($R_S = 1.5$) can be selected if better separation is required. On the other hand, as R_S becomes less than 1 the bands overlap more and more. When R_S is less than 0.8, the separation is usually unsatisfactory.

An interesting point can be made in terms of the effect of migration distance on the peak maxima separation and the band width. The peak maxima separation ($t_{R2} - t_{R1}$) is directly proportional to the distance of migration. However, the peaks are broadened only with the square root of distance of migration. Therefore, the bands separate faster than they broaden in a column, so that, as is well known, a longer column gives a better separation. This omits the fact that a longer column represents a longer retention time (see equation 1.6).

---

*It can be argued that considering the resolution of only a two-component system is unrealistic when one deals with a multicomponent mixture (ca. sixty species). For such a mixture a picture would be worth a thousand words. However, consideration of a two-component mixture can be justified even in this case, for one pair of the sixty components in the mixture may be the most difficult to separate and therefore would be of the greatest concern.

It must be noted, moreover, that ultimately a separation
is judged on the resolution and the analysis time.

Equations 1.7 and 1.8, although useful in measuring
the degree of separation, do not relate the fundamental
chromatographic parameters to resolution. If we assume
equal band widths (closely spaced peaks), a relationship
can be derived between resolution and three fundamental
chromatographic parameters (4):

$$R_s = \frac{1}{4} \left(\frac{\alpha - 1}{\alpha} \right) \left(\frac{k_2'}{1 + k_2'} \right) (N_2)^{1/2} \qquad (1.9)$$

where α is equal to the relative retention, N is the number of theoretical plates, and the subscript 2 represents
the second component. Equation 1.9 is probably the most
important equation in chromatography and consequently is
well worth examining in detail. Let us look at each
term individually.

The relative retention, α, is defined in the equation

$$\alpha = \frac{t_{R2} - t_o}{t_{R1} - t_o} = \frac{k_2'}{k_1'} = \frac{K_2}{K_1} \qquad (1.10)$$

It can be seen that α is the net retention time ratio
for the two components and can be simply calculated from
the chromatogram. More fundamentally, α is equal to the
ratio of equilibrium distribution coefficients or capacity factors for the two components. In effect, α
relates to the peak-to-peak separation of the two components and is a measure of the thermodynamic differences
in their distributions:

$$\Delta(\Delta G^o) = -RT \ln \alpha \qquad (1.11)$$

where $\Delta(\Delta G^o)$ is the difference in free energies of distribution for the two components.

When $\alpha = 1$, the resolution is zero (equation 1.9),
regardless of the number of theoretical plates in the
column. There must be some difference in the thermodynamic distribution behavior of the two components for
separation to occur. Separations for α values as low
as 1.01, however, can be achieved in some forms of chromatography. In addition, from the functionality in

equation 1.9, small changes in α can mean large changes
in R_S, especially when α is close to 1. Thus a change
in α from 1.1 to 1.2 (a 10% change) results in approximately a doubling in resolution through the $(\alpha - 1)$ term.

The relative retention relates to the differences in
interactions of the two components in the mobile and stationary phases. The better the forces of intermolecular
interactions are understood, the clearer will be the com-
prehension of selectivity in chromatography. Consider
as a simple example the separation of a ketone from an
alcohol. If LLC is employed as a means of separation
and an electron-donating solvent is used as a stationary
phase, the alcohol will be retarded relative to the ketone via hydrogen bond formation with the solvent. The
separation then becomes a simple one, requiring very few
theoretical plates.

The better the selectivity of the chromatographic system, the easier will be the separation. From this point
of view, liquid chromatography has one big advantage
over gas chromatography. In the latter, interactions
in the gas phase are negligible, and thus only the stationary phase can be used for creating thermodynamic
differences of distribution (along with vapor pressure
differences). In liquid chromatography, on the other
hand, the mobile phase is no longer inert, but can play
a central role in the thermodynamic distribution process
through selective interactions in the mobile phase. In
LSC the mobile phase is in selective competition with
solute molecules for adsorption sites on the adsorbent.
This is the major reason why classical LC has been so
successful. Even though the columns are inefficient,
the α values are sufficiently high that separation can
be achieved. The combination of this advantage of potentially high α values with efficiencies and speeds comparable to those obtained in GC makes high-speed LC a most
powerful separation tool. In Chapter 4, the role of
solvents in selectively invoking the forces of intermolecular interactions is discussed in more detail.

The second parameter in the resolution equation is N,

the number of theoretical plates,*

$$N = 16 \; (\frac{t_R}{w})^2 = (\frac{t_R}{\sigma_t})^2 \qquad (1.12)$$

where σ_t equals the standard deviation of the Gaussian function in time units. The number of theoretical plates is related to the relative band broadening in the chromatographic column. The term relative is used because, as already noted, a band automatically broadens the longer the time of migration. However, the extent of band broadening after a certain time of migration is the parameter of interest. In effect, N is related to the column efficiency.

In equation 1.9 the resolution is proportional to the square root of the number of theoretical plates. This relationship has important consequences for the achievement of a separation in chromatography. For example, from equation 1.12 it can be predicted that N is proportional to column length (t_R is proportional to L, and w to $\sqrt{L}$). Thus doubling the column length will increase the resolution by a factor of 1.4. A price is paid for this improved resolution, however, and this price is an increase in time. Equation 1.6 shows that the retention time is directly proportional to the column length. Thus doubling the length of the column increases the time of separation by a factor of 2. There are, however, other parameters which can be changed to improve separation without such a concomitant increase in retention time. An understanding of these parameters has made high-speed LC an important and practical tool.

The third parameter in equation 1.9 which determines resolution is the capacity factor for the second component, k_2'. The capacity factor has already been discussed in relation to retention, and here its importance in

---

*Actually the number of theoretical plates for the second component appears in equation 1.9. For this discussion the simplifying assumption will be made that $N_1 = N_2$.

resolution is seen. When $k_2' = 0$, $R_s = 0$, and both components come out with the nonsorbed time. The resolution
improves as k_2' increases until, at large values of k_2',
the k' function goes to unity and therefore no longer
plays a role in resolution. In classical packed GC columns the capacity factor is often sufficiently high
(greater than 20) that the term is neglected. However,
in high-speed LC (as well as capillary GC) the capacity
factor is small and must be considered. The small k'
value arises from the fact that the volume ratio of stationary to mobile phase is small. Small k' values are
desired for high speed, and high k' values for resolution. A compromise must obviously be made, that is,
there is an optimal k' value. This will be discussed
later.

Many workers prefer to simplify the resolution equation 1.9 by introducing the number of effective plates in
the column, defined as

$$N_{eff} = (\frac{k'}{1 + k'})^2 (N) = 16(\frac{t_R - t_o}{w})^2 = (\frac{t_R - t_o}{\sigma_t})^2 \qquad (1.13)$$

The number of effective plates in a column gives more
clearly the true separating efficiency of the column;
for example, even if $N = 100,000$ in a column, k' can be
so small that a sufficient number of effective plates
are not available to achieve the separation (see 5).
Upon substituting with N_{eff} the resolution equation
becomes

$$R_s = \frac{1}{4} (\frac{\alpha - 1}{\alpha}) (N_{eff})^{1/2} \qquad (1.14)$$

This equation is simpler than the previous resolution
equation and divides separation into two approximately
independent factors. If a resolution of 1 is assumed,
the functionality of α versus N_{eff} can be plotted as in
Figure 1.4. Here, when α is very small, a large number
of effective plates are needed, and as α increases the
number of effective plates required decreases. The
slope of the curve is such that, as previously noted,
small changes in α when α is close to 1 result in large
decreases in the N_{eff} required. Since the required

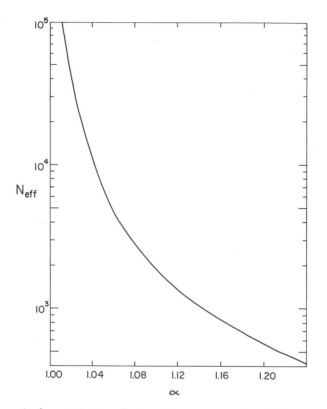

Figure 1.4. Plot of log N_{eff} required for achievement of a resolution = 1 as a function of relative retention, α.

number of effective plates is roughly proportional to time, a larger α value means a shorter separation time.
    Table 1.1 shows the number of effective plates required to achieve a given resolution, R_s, with a given α value. Note the larger number of effective plates required for better resolution. From equation 1.14 it becomes a simple matter to predict whether a separation is possible (using a certain column and conditions) for a given α value. As a rule of thumb, 500-750 effective plates constitutes a reasonable range of values for the solution of most of the separation problems encountered in liquid chromatography.

Table 1.1

Number of Effective Theoretical Plates
Required for Separation as a Function of α

$R_S = 1$		$R_S = 1.5$	
α	N_{eff}	α	N_{eff}
1.01	160,000	1.01	360,000
1.05	6,800	1.05	15,700
1.10	1,940	1.10	4,360
1.15	940	1.15	2,110
1.2	575	1.2	810

Finally, Figure 1.5 presents three chromatograms to illustrate the importance of the chromatographic parameters in regard to resolution. In the top figure, the peaks are relatively sharp and reasonably well removed from the nonsorbed time. However, the relative retention, α, is poor and separation does not result. Condition changes to increase α would most easily improve resolution. The middle chromatogram also shows an incomplete separation, due to small k' values (and a low N_{eff}). Here, the increase in the stationary-mobile-phase volume (area) ratio or an increase of k' by changing temperature would improve resolution. The bottom chromatogram illustrates a multicomponent separation in which all the bands are not resolved. In this case an increase in the number of theoretical plates in the column may be the best approach to better separation because N is a general parameter -- an increase in N (e.g., longer columns) will sharpen all bands in respect to each other. On the other hand, a change in conditions to increase α (e.g., a change in the stationary phase) may resolve one pair of components while overlapping a new pair. Multicomponent analysis will be discussed in Section C.4.

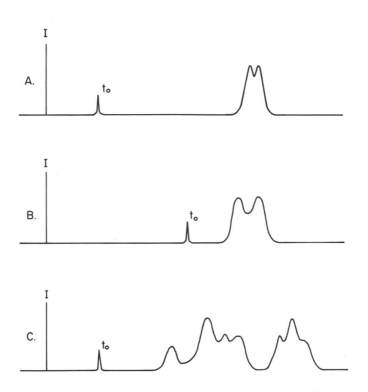

Figure 1.5. Chromatograms showing the influence of α (A), k' (B), and N (C) on resolution.

2. Height Equivalent to a Theoretical Plate

As noted, an optimum separation is one in which adequate resolution occurs in the minimum amount of time. Before proceeding to a discussion of optimization, it is first necessary to examine the relationship between the height equivalent to a theoretical plate, H, and mobile-phase velocity. Height H is defined as

$$H = \frac{L}{N} = \frac{\sigma_L^2}{L} \qquad (1.15)$$

where σ_L is the standard deviation of the Gaussian function in length units. Equation 1.15 assumes a uniform column and compressible fluid. Since a large number of theoretical plates are desired, H should be as small as possible. The importance of the variation of H with v in obtaining separation in a minimum time can be seen from a substitution of equation 1.15 into equation 1.16:

$$t_R = N(1 + k')\frac{H}{v} \qquad (1.16)$$

The value of N in equation 1.16 will be dictated by the demands on resolution and the α value. The time of analysis, t_R, is seen to be directly proportional to H/v.

In equation 1.15, H is proportional to the variance of the Gaussian function per unit column length. If independent processes in the column (extra column effects are neglected here) are assumed, with each process contributing a certain variance to band broadening, then the total variance of equation 1.15 can be obtained by summing up the individual variances. Each of the processes causing band broadening will be examined, and a variance will be attributed to each. The practical conclusions of these relationships will be emphasized in this discussion.

Before proceeding with this development, it should be noted that some workers have preferred to use reduced parameters for H and v (5). The reduced plate height, h, is equal to H/d_p, where d_p is the particle diameter of the support. The reduced velocity, ν, is equal to vd_p/D_M, where D_M is the diffusion coefficient in the mobile phase. The reduced plate height normalizes H for particle diameter, and the reduced velocity normalizes for diffusion over a particle diameter distance. This normalization makes it possible to compare a wide variety of chromatographic systems.

In this chapter, the more practical quantities H and v are preferred because they can be directly related to optimization through equation 1.16. In addition, as noted in Chapter 9, erroneous conclusions can sometimes be drawn when reduced plate heights are compared for different chromatographic conditions, if H is not proportional to d_p^2. It may further be noted that in gas

chromatography the actual mobile-phase velocity for a
packed column is about 2-15 cm/sec. In high-speed
liquid chromatography the velocity ranges between 0.1
and 5-10 cm/sec. However, because the diffusion coefficient is very much smaller in liquids than in gases, the
reduced velocity is much larger in liquid than in gas
chromatography (e.g., $10-10^3$ for LC versus 0.1-5 for GC).
It is difficult, therefore, to compare gas and liquid
chromatography at the same reduced velocity on a practical basis.*

Figure 1.6 shows a plot of H versus v for a retained
solute in gas and in liquid chromatography.** The shapes
are distinctly different, the major reason being that

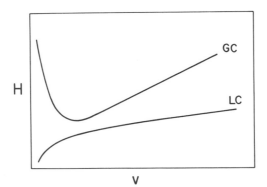

Figure 1.6. Typical H versus velocity curves for
gas and liquid chromatography.

diffusion coefficients in liquids are 10^4-10^5 times
smaller than those in gases. If the GC plot is examined

---

*The decided theoretical advantages of using the reduced
parameters are recognized. However, in the context of
this chapter, that is, the relationship of theory to
practice, H and v are more useful parameters.
**The two plots are displaced from one another for clar-
ity; however, they may actually overlap to some extent,
depending on the particle diameters of the supports used
in the two methods.

first, it is noted that the shape of the curve is a hy-
perbola, for which a simple equation can be written:

$$H = A + \frac{B}{v} + Cv \tag{1.17}$$

where no correction is made for gas compressibility.
Band-broadening processes will now be attributed to each
of the terms--A, B, and C.

The B term relates to band broadening occurring by
diffusion in the gas phase in the longitudinal direction
of the column. The one-dimensional Einstein equation
for diffusion is (6)

$$\sigma_L^2 = 2D_M t_o = \frac{2\ D_M L}{v} \tag{1.18}$$

and, since $H = \sigma_L^2/L$, it is readily seen that $B = 2\ D_M$.*
The inverse velocity function results in this term being
important at low velocities (i.e., below the minimum H
in Figure 1.6). Since D_M is ca. 10^5 times smaller in
liquids than in gases, this longitudinal term plays no
practical role in band broadening in LC. Consequently,
with this technique no decrease in H with increasing
velocity is observed at low velocities, and indeed a minimum H is rarely measured (see the plot of H versus v
for LC in Figure 1.6).

The A term in equation 1.17 arises from inhomogeneous
flow processes (eddy diffusion). In an oversimplified
picture, this term can be seen to arise in the following
manner. The path lengths of the various channels in the
packed bed are different, and as a consequence solute
molecules reach the detector at different times. In
this case, the variance is

$$\frac{\sigma_L^2}{L} = 2\lambda d_p = A \tag{1.19}$$

where λ is a packing correction factor of ~ 0.5. In

---

*Actually, a correction factor, $\gamma = 0.7$, is included to
take account of the tortuous diffusion path in a packed
bed.

classical GC the A term is a constant, representing a
lower limit on column efficiency, that is, H = d_p or h
= 1. Of course, this lower limit is not reached in practice, but with high-efficiency columns $H_{min} \cong 3 - 5$ d_p
can be achieved. The A term occurs also in LC columns;
however, as will be seen, it appears in a somewhat different form.

At velocities above H_{min}, the C term controls H; and
since high velocities are obviously necessary for high-
speed analysis, equation 1.16 can be written for GC as

$$t_R = N(1 + k')C \qquad (1.20)$$

The C term relates to nonequilibrium resulting from resistance to mass transfer in the stationary and mobile
phases. A number of forms of resistance to mass transfer can exist (7) (e.g., slow diffusion in a stationary
liquid phase, slow adsorption or desorption from a
surface).

Often workers do not understand or appreciate the
critical importance of these mass transfer effects on
the speed of analysis, and so it is worthwhile to present a brief picture of how these effects result in
band broadening. The arguments presented by Giddings
(8) will be followed. Figure 1.7 illustrates the
equilibrium-concentration zone profiles (full lines)
and the actual-concentration zone profiles (dashed
lines) in both the stationary and the mobile phases for
a given solute. Note that the actual concentration
leads the equilibrium zone in the mobile phase and
trails the equilibrium zone in the stationary phase.
The displacements are equal on both sides, resulting in
a symmetrical broadening of the zone in its travel down
the column.

The extent of broadening (or nonequilibrium) is a
direct function of the rate of flow relative to the rate
of mass transfer. This can be seen by examining the
processes that occur as the mobile phase moves down the
column. As the concentration zone moves into a region
that is devoid of solute, transfer of solute from the

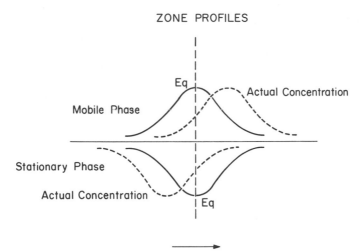

ZONE PROFILES

Figure 1.7. Illustration of the influence of local nonequilibrium on band broadening. Full lines, equilibrium-concentration zone profile; dashed lines, actual-concentration zone profile.

mobile phase to the stationary phase takes place in an attempt to reach equilibrium. However, as the flow proceeds the concentration in the mobile phase continues to increase in the given region (until the peak maximum is reached), and hence equilibrium is not attained. The extent of local nonequilibrium is clearly a function of the rate at which solute flows into the region relative to the rate at which transfer from the mobile to the stationary phase occurs. Analogous arguments can be made for the back side of the Gaussian peak. Slow flow and rapid mass transfer signify that local nonequilibrium will be slight and column efficiency will be good. Since rapid velocities are required for high-speed analysis, it is critical that rates of mass transfer be rapid.

Rate-limiting processes for mass transfer that determine the extent of local nonequilibrium in gas chromatography include (a) diffusion of solute in the mobile phase, (b) diffusion in the stationary liquid phase

(GLC), and (c) adsorption-desorption kinetics (GSC). In
liquid chromatography processes (b) and (c) are equivalent to their GC counterparts (in terms of the mathemati-
cal formulation, but not necessarily in terms of relative
importance); however, mobile-phase diffusional processes
(a) are more complex. The contributions of each of these
terms to H will now be examined for GC. The correspondence of processes (b) and (c) in GC and LC will simplify
the later discussion of mass transfer effects in the
latter technique.

For both processes (b) and (c) a simplified general
expression can be derived from the random walk model:

$$H = 2 \ \frac{k'}{(1 + k')^2} \ vt_d \qquad (1.21)$$

where t_d is the mean desorption time (the average time
that the solute molecule spends in the stationary phase
between transfers to the mobile phase).* First, note
that H is proportional to v; a faster velocity means less
chance for the attainment of equilibrium. Second, the
capacity factor function in equation 1.21 reaches a maximum at $k' = 1$ and then decreases with increasing k'
values, the function being approximated as $1/k'$ for large
k' values. From a qualitative standpoint, it is expected
that H should decrease with increasing k' for stationary-
phase mass transfer effects, because a large k' value
means a slow flow of solute through a given region of the
column.** Finally, equation 1.21 shows the direct proportionality of H and t_d, indicating that the rate of
mass transfer should be as rapid as possible. In desorption kinetics, $t_d = 1/k_d$, where k_d is the desorption rate
constant. (Here a simple one-site surface for adsorption

---

*For rate-limiting processes in the mobile phase, H =
$2[k'/(1 + k')]^2 \ vt_a$, where t_a is the sorption time.
**Do not confuse mobile-phase velocity, v, with solute
migration velocity, $v/(1 + k')$.

is assumed. The equation must be modified for more complex adsorption processes.)
    For stationary liquid-phase diffusion, the Einstein
equation 1.18 can be solved to give $t_d = d^2/2D_S$, where d
is the diffusion path length and D_S is the stationary-
phase diffusion coefficient. Equation 1.21 then becomes

$$ H = \frac{k'}{(1 + k')^2} \frac{d^2 v}{D_S} \qquad (1.22) $$

Here d can be equated approximately with the thickness
of the liquid layer (an exact correspondence is not
found), so that the greater the liquid loading the
slower will be the rate of mass transfer. A coefficient,
q, is placed on the right side of equation 1.22 in a
more exact formulation to take account of the different
pore shapes in which the liquid can coat the support
(see ref. 7).
    Diffusion is more complicated in the mobile than in
the stationary phase, and a detailed discussion of this
phenomenon is beyond the scope of this chapter. However,
a brief summary as to how the term arises will be helpful. Figure 1.8 illustrates the flow profile in an open

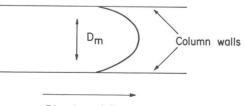

Direction of Flow

Figure 1.8. Flow profile in an open tube; D_M =
radial diffusion coefficient to relax profile.

tube in which the velocity is fastest in the center and
decreases to the column walls. This flow profile will
produce a concentration profile, and a solute at the
center of the profile will reach the detector before the

material at the walls. To relax* this distribution, it
is necessary for radial diffusion (or mixing) to occur
(i.e., diffusion in a lateral direction to that of the
flow). In this manner solute molecules move from one
flow zone to another, and the velocities of all solute
molecules become more nearly equal. When the column is
packed with particles, the flow pattern becomes significantly more complex; however, the same principle of
broadening occurs. Although relaxation of the concentration profile again occurs by the radial diffusion
process, convection (due to the interruption of the flow
pattern by the intervening particles) now also plays a
significant role. Giddings (8) has derived the term
relating to this process as

$$H = \frac{\omega d_p^2 v}{D_M} \qquad (1.23)$$

where ω is a dimensionless constant, and d_p is the particle diameter.** Since ω has not been derived formally
to date, equation 1.23 is an approximation to diffusion
in the mobile phase. Note in this equation the indepen-
dence of k' and the fact that $H \sim d_p^2$. The latter
dependence means that the C term is smaller, the smaller
the particle diameter.

Now that the general band-broadening effects have
been discussed in terms of gas chromatography, liquid
chromatography will be considered. Figure 1.6 shows
a markedly different shape of the H versus velocity curve
for the latter technique. An equation for this curve
can be written as follows (7):

---

*Relaxation of the profile will result in equalization
of the concentration of the sample in the tube and a
smaller band width.
**In a more complicated version of equation 1.23, five
velocity inequalities are considered (e.g., across the
particle, across a channel, across the column). Each
contributes ω_i to the band-broadening term, so that $\omega = \sum_i \omega_i$.

$$H = C_S v + C_M^* v + (\frac{1}{A} + \frac{1}{C_M v})^{-1} \qquad (1.24)$$

The $C_S v$ term represents stationary-phase mass transfer effects similar to those described for GC. The $C_M^* v$ term represents slow mass transfer caused by solute diffusion in the mobile phase, which is stagnant in the pores of the support material.* The slow diffusion in liquids makes this term necessary in equation 1.24. Equation 1.25 accounts for this effect (8):

$$H_M^* = C_M^* v = \frac{1}{30} f(\Phi', k') \frac{d_p^2 v}{D_M} \qquad (1.25)$$

where Φ' = the fraction of mobile phase occupying intra-particle space. Note the dependence of this mobile-phase transfer term on k'. In addition, this term decreases, the smaller the particle diameter and the smaller the pore depths filled with mobile phase. More simply, the smaller the path length for diffusion, the more rapid is the mass transfer. Specially designed particles have been used to overcome the importance of this term, namely, porous layer beads. These will be described in Section D.1.

Figure 1.6 shows a much flatter rise of H with velocity in liquid than in gas chromatography. This is a result of very complex flow processes occurring in the liquid mobile phase. Since diffusion is slow, convective lateral mixing plays a very significant role in the relaxation of the velocity profile. According to Giddings (9), this complex flow pattern results in a coupling of the eddy diffusion and mobile-phase terms to give

$$H = (\frac{1}{2\lambda d_p} + \frac{D_M}{d_p^2 v})^{-1} \qquad (1.26)$$

---

*Some workers differentiate the two types of mobile phase as the "stagnant mobile phase" and the "moving mobile phase."

According to this equation, H should be proportional to
some power of d_p less than 2. Also, at high velocities
H should converge to the classical eddy diffusion term.
Note again the absence of k' in equation 1.26. For a
detailed discussion of this term see the series of
papers by Knox (10).

In many LC columns, mobile-phase effects predominate
over stationary-phase effects. If in addition the contribution to band broadening from diffusion in the stagnant mobile phase (equation 1.25) is small, H will be
independent of k' (27). At other times, when this broadening process is important, H is found to be dependent
on k' even if stationary-phase mass transfer effects are
absent (10). In addition, the flat slope of the H versus v plot means that high mobile-phase velocities can
be used without excessive loss in column efficiency.
After a discussion of the principles of time optimization in the next section, the features of liquid chromatography that are incorporated into high-speed separations
will be described in detail. These features are pointed
out in the band-broadening equation 1.24.

This discussion of band broadening will be concluded
with a brief examination of the influence of extra-
column effects on column efficiency. Band broadening
outside the column will contribute a variance to H which
in general will decrease apparent column efficiency and
result in a loss in resolution. Loss in efficiency can
result from poor injection, broadening in the tubing
between the column and the detector, and from within the
detector itself. On-column injection or injection into
glass beads at the top of the column is a recommended
procedure (11). Although a sampling valve can be used,
one must be cautious of solute being slowly removed from
the walls of the valve. Injection is especially critical in regard to potential loss in column efficiency.
Poor injections particularly produce inferior separations of solutes having small k' values.

In liquid chromatography, if good injection procedures are assumed, a major influence on efficiency can
occur in the connection between the column and the

detector ($\underline{12}$). Indeed, this effect is much more pronounced in liquid than in gas chromatography because of
the lower diffusion coefficients in liquids. For this
reason it is very important that the volume of the connector be as small as possible. Obviously, the dead
volume in the detector must also be minimized; this can
be a major source of extra-column band broadening if
precautions are not taken. Finally, when an ultraviolet
photometric detector is used, the design of the cell
through which the sample flows can be important ($\underline{13}$).

Extra-column effects add a variance to the column
variance:

$$\sigma_T{}^2 = \sigma_{col}^2 + \sigma_{ex-col}^2 \qquad (1.27)$$

Since the column and the tube connection are probably of
different diameters, the volume variances should be used
in this equation. Now the influence of σ_{ex-col}^2 will be
directly tied to the value of σ_{col}^2. Clearly, the more
efficient the column (the smaller σ_{col}^2), the greater the
need to design the system with low dead-volume connec-
tions, good injectors, etc. In classical LC there is
little concern with extra-column effects, but in the
high-speed technique they play a most important role.

For a given column, the influence of extra-column
effects depends on the k' value of the solute. As k'
increases, σ_{col}^2 also increases and the relative importance of σ_{ex-col}^2 decreases. The critical test, then,
for the influence of the equipment on the apparent efficiency of a given column is the form of the H versus
v curve for an unretained peak. If the form of the
curve for the nonsorbed species is unusual ($\underline{12}$), or if
H at a given velocity is unusually high (relative to H
for solutes with k' values over 1-2), then extra-
column effects are probably important. Column variance,
σ_{col}^2, depends also on column length (the longer a component is in the column, the broader the band), so that
extra-column effects become more important the shorter
the column. It is possible, therefore, to determine the
variance of extra-column factors directly by measuring

H for an unretained component as a function of column
length at a given velocity and extrapolating to zero
length. For a complete discussion of extra-column
effects in chromatography, see ref. 14.

Finally, in the discussions of R_S and H linear elution
conditions in which symmetrical Gaussian peaks are obtained have been assumed. In high-speed LC the detector
sensitivity is such that sufficiently small sample sizes
can be injected for isotherm linearity. However, if a
low-capacity solid support is used, it is possible that
an ordinary sample size may overload the column, resulting in band asymmetry. A decrease in sample size should
overcome this problem. Band asymmetry (tailing) can also
arise from a poor injection in which the sample is carried on to the column, not as a plug but as an exponential zone. Peak asymmetry is undesirable for two
reasons: first, the bands are broadened with a loss in
resolution; and, second, the peak maximum (retention
time) becomes a function of the amount of sample injected
if one operates in the nonlinear region of the isotherm.

3. Optimization and Time

Resolution and separation time will now be discussed.
In effect, it is desired that R_S/t_{R2} be large. This can
be achieved by keeping R_S constant and finding conditions
to make t_{R2} as small as possible (minimum time analysis),
or by keeping t_{R2} constant and finding conditions to make
R_S as large as possible (time normalization). Minimum
time analysis, defining $R_S = 1$ as adequate resolution,
will be examined here. (When time is the important
parameter, any resolution greater than this can be detrimental to the speed of analysis.)

As already noted, the retention time can be written as

$$t_R = N(1 + k') \frac{H}{v} \qquad (1.16)$$

Now, with $R_S = 1$, N will represent the number of plates
required to achieve the separation. The larger the value of α, the smaller will be N and, from equation 1.16,
the shorter the analysis time. Thus, the first requirement when trying to perform a separation is to select

phases which maximize α. For the following discussion
it will be assumed that a good stationary phase for a
given pair of solutes has been selected. This value of
α will be constant, and the other parameters which permit t_R to be small will be examined.

Figure 1.9 shows a typical plot of H/v versus v to
illustrate the influence of v on analysis time. Note
that H/v drops sharply at first with increasing velocity,

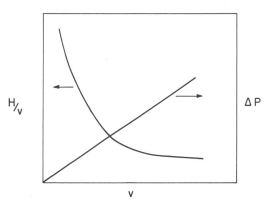

Figure 1.9. Plots of H/v versus v and ΔP versus
v to illustrate the influence of velocity on time
and pressure.

followed by leveling off. In gas chromatography, where
H = Cv at velocities somewhat greater than v_{opt}, H/v becomes a constant. In liquid chromatography, because of
the coupling of the A and the C_M terms, H/v continues to
decrease with increasing velocity; however, the slope is
small. Experimental confirmation of this behavior in an
ion-exchange column can be found in ref. 15. The velocity at which the slope of the H/v versus v plot stops
changing significantly can be considered as an optimum
velocity, v'_{opt}, for rapid analysis. In GC, this velocity
is found to be roughly twice v_{opt}, whereas in LC it is
less clearly defined but is probably close to 2-3 cm/sec
for a 2-3 mm i.d. column (16,17).

At first glance it may appear worthwhile to continue
to increase velocity in LC beyond v'_{opt} since H/v slowly
decreases. A price is exacted, however, for this

increased speed, namely, pressure drop in the column, ΔP. Pressure drop is related to velocity through the column permeability, K^O:

$$K^O = \frac{v\eta Lf}{\Delta P} \qquad (1.28)$$

where η is the mobile-phase viscosity and f is the total porosity in the column.* Pressure is a critical parameter in LC, and in many cases can be the limiting parameter affecting the ultimate speed and separation capabilities of a column.

Since the viscosity of liquids is 100 times greater than that of gases, the pressure drop will be 100 times greater in liquid than in gas chromatography for a given L, v, and K^O. In addition, from the Kozeny-Carman equation (1.18) for a regular packed column,

$$K^O = (\frac{d_p^2}{180}) \left[\frac{\epsilon^3}{(1 - \epsilon)^2} \right] \qquad (1.29)$$

where ϵ is the interparticle porosity.** For a regular packed column $\epsilon = 0.42$, so that a good approximation of equation 1.29 is:*** (5)

$$K^O = \frac{d_p^2}{1000} \qquad (1.30)$$

As already noted, particle diameters smaller by as much as a factor of 6 are used in liquid relative to gas

---

*This equation is valid for a regular packed column.
**Note that f and ϵ are not necessarily the same. For a column packed with glass spheres, ϵ = f. For porous beads, $\epsilon \simeq 1/2F$, with $f \sim 0.84$.
**Equations 1.28 and 1.30 are very useful as diagnostic tools in examining an LC system. Thus K^O can be estimated from the particle diameter of the support, and then the pressure drop to achieve a given velocity for a given column length can be calculated. If the predicted and the experimental pressure drops differ by more than 20%, is is probable that a leak or a plug exists in the system. Actually, in the case of a plug, the experimental and the calculated ΔP's usually differ by a factor of 2 or more.

chromatography (the reason will be examined shortly).
Thus, the pressure drop is again seen to be much higher
in LC: typical pressures are 20-40 psi in GC but 500-
3000 psi in high-speed LC. Of course, there is little
danger in using these pressures, since liquids are incompressible; however, high-pressure pumps are required.

The importance of pressure and permeability in regard
to analysis time can be seen by substituting v from
equation 1.28 into the retention time equation 1.16:

$$t_R = \frac{\eta L^2 f(1 + k')}{K^o\, \Delta P} = \frac{1000\eta f(1 + k')}{\Delta P}\, (\frac{L}{d_p})^2 \qquad (1.31)$$

For a constant $L, k', \eta, f,$ and ΔP it is seen that the more
permeable the column, the shorter is the analysis time.
Concurrently, if t_R is maintained constant (time normalization), a more permeable column will permit the use of
a longer column, resulting in better resolution (more
theoretical plates) and a larger peak capacity (see the
next section).

From equation 1.30, the permeability may be increased
by increasing the particle diameter; however, larger
particles mean poorer H values. In fact, as will be seen,
the advantage of higher pemreability is more than offset
in this case. Equation 1.30 is based on a regular packed
column for which the ratio of column diameter, d_c, to
particle diameter is greater than 10. As shown by
Halasz (19) and Knox (20), when d_c/d_p is less than 5, H
decreases and the permeability increases by a factor of
10 or more. The increase in efficiency arises from the
increased radial mixing possible in the mobile phase.
In liquid chromatography such irregular packed columns
would seem to offer real advantages over regular packed
columns in that more than a factor of 10 decrease in t_R
or increase in L could be achieved for a given pressure
drop. However, up to now such columns have been difficult to make reproducibly.

With regular packed columns, equation 1.31 can be
used to understand the interrelationship of L/d_p with t_R
and ΔP. The larger the value of L/d_p, the more theoretical plates will be available for separation. Increases

in L/d_p must be bought at the price of either increased
t_R or ΔP. Undoubtedly, it is preferable in most cases
to increase pressure; however, an upper pressure limit,
ΔP_{lim}, is imposed by the equipment. When this pressure
is reached, increased efficiency can be attained only
with concomitant increases in retention. At constant
L/d_p ratio, a column of small L and d_p will produce a
faster analysis than one of longer length and larger
particle size. In this regard, Knox (21) has recommended
a column of 10-cm length and 1 μ particle diameter as
the ideal configuration; however, the technical problems
with developing small-volume injection ports and detectors are formidable. In addition, small particles (less
than 20 μ) are very difficult to pack reproducibly.
More will be said on this point when specific conditions
for high-speed liquid chromatography are discussed in
Section D.

As noted earlier, it is useful to consider effective
plates as the critical efficiency parameter for separation. If time is incorporated into the separation problem, it is also useful to introduce a new term -
effective plates per second, N_{eff}/t. Since a given number of effective plates are required for a separation
(see Table 1.1), N_{eff}/t can tell directly how long the
separation must take. Combining equations 1.16 and 1.13
gives

$$\frac{N_{eff}}{t} = \frac{v}{H} \frac{(k')^2}{(1 + k')^3} \tag{1.32}$$

A maximum N_{eff}/t means the shortest analysis time.

In Figure 1.10 the k' function in equation 1.32 is
plotted versus k'. If H is assumed to be independent
of k' (as will be the case when mass transfer in the
mobile phase is the predominant band-broadening mechanism), then the k' function is directly related to
N_{eff}/t. The optimum k' value is seen to be 2. Higher
k' values result in longer retention; lower k' values,
in a lower N_{eff}. When H is allowed to be a function of
k' (important contributions to band broadening from
stationary-phase mass transfer processes or from stagnant

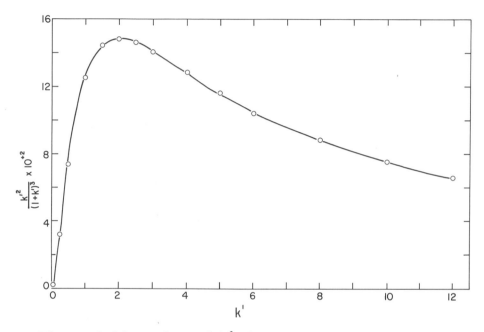

Figure 1.10. Plot of k' function in N_{eff}/t equation versus k'. Note the maximum at k' = 2.

mobile-phase diffusion), the optimum k' value occurs anywhere between 1.5 and 4.

Figure 1.10 is informative also in other ways. First it is seen that the maximum is fairly broad and that the ordinate does not change greatly between k' = 1.5 and 3. Therefore, working exactly at k' = 2 is not necessary. Second, k' values below 1.5 are very detrimental to the speed of analysis. Thus, for example, N_{eff}/t drops by a factor of 2 when k' is changed from 1 to 0.5. On the other hand, there is a considerably slower decrease in N_{eff}/t when k' is increased over 2.

4. Peak Capacity

Up to now the separation of a two-component system has been discussed, and it has been suggested that the same basic principles can be extended to multicomponent separations. The importance of the number of theoretical plates in multicomponent analysis has been mentioned. This section shows more specifically how this importance arises.

To help quantify a discussion of multicomponent analysis, Giddings introduced the term peak capacity (22). If a constant number of theoretical plates (independent of k') is assumed, the peak capacity can be defined as the number of peaks that can be placed within a certain time period in which all of the bands have a resolution of unity. The first band elutes at the nonsorbed time, and the final band has its peak maximum at the time specified as the end of the chromatogram.* Mathematically, if a 4σ separation is assumed, the peak capacity, $\emptyset$, can be written as

$$\emptyset = 1 + 0.6N^{1/2}\log (1 + k') \qquad (1.33)$$

where k' is the capacity factor for the final component. This equation is plotted in Figure 1.11 in terms of log (1 + k') versus peak capacity. It is noted that the slope of the line is a function of the square root of the number of theoretical plates in the column. Therefore, for a given capacity factor for the final component, an increase in the number of theoretical plates increases the peak capacity. In addition, when there is insufficient peak capacity for a given separation, Figure 1.11 shows that an increase in k' will improve $\emptyset$ for the column with the larger number of theoretical plates.

---

*The small increment of time necessary for the final peak to return to the base line is neglected.

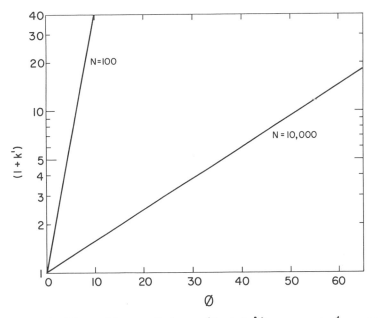

<u>Figure 1.11.</u> Plot of log (1 + k') versus ∅, peak
capacity for two columns of different efficiencies.

On the basis of peak capacity, gel permeation chromatography is seen to have a considerable disadvantage
relative to other methods. This arises from the fact
that the capacity factor can reach a value of only about
2.5-3. In gas chromatography, on the other hand, the
capacity factor can be of the order of 50 or 100 or even
higher. In high-speed liquid chromatography using porous layer beads the capacity is again small; however, it
can be increased by means of a more favorable thermodynamic distribution. As a result the peak capacity is
significantly less in GPC than in gas or liquid chromatography.

Horvath and Lipsky (<u>23</u>) have noted that the peak
capacity for a gas or liquid chromatographic system can
be significantly improved through the use of one or more
gradient techniques, such as gradient elution or temperature programming. This arises from the fact that
all of the peaks elute with the same width, and

therefore the chromatographically calculated number of
theoretical plates increases with increasing k'. Thus
the number of plates required to resolve 100 components
in regular elution chromatography might be 25,000, compared to perhaps 700 with gradient elution. Gradient
elution techniques are used not only for achieving peak
capacity but also for overcoming the general elution
problems, namely, the bunching of components at the beginning of a chromatogram and the long time required for
the elution for later peaks. In Chapter 6, gradient
elution techniques are discussed in detail.

D. HIGH-SPEED LIQUID CHROMATOGRAPHY

The fundamentals of separation and optimization in
chromatography have now been examined. Next, the developments in high-speed liquid chromatography which can be
understood in terms of these fundamentals will be
described.

1. Porous Layer Beads

In most forms of classical liquid chromatography,
porous packing materials, such as silica, alumina, and
ion-exchange beads, are used. In adsorption (including
ion-exchange) chromatography such materials are useful
for the high surface area available for adsorption.
However, they become disadvantageous when one desires
to perform high-speed liquid chromatography. The mass
transfer term due to diffusion in the stagnant mobile-
phase in the pores of the support becomes important,
limiting the speed attainable (with reasonable efficiency). Clearly, for high-speed analysis, the migration
distance or pore depth must be decreased.

Golay (24) and Purnell (25) first suggested from a
theoretical point of view that support materials composed
of thin porous layers of adsorbent surrounding solid
impenetrable cores (i.e., porous layer beads, PLB) could
be useful as a packing material for rapid analysis.*

---

*Knox and McLaren (43) have also pointed out the advan-
tages of PLB for high-speed chromatography.

Halasz and Horvath (26) were the first to prove the success of these materials in GC. Many of the new support materials used in high-speed LC are basically, PLB, namely Zipax® (27), pellicular ion-exchange resin beads (28), surface-etched beads (2), and Corasil® I and II (29). To give good packing density, these beads are spherical in shape. In addition, they have the big advantage (relative to ion-exchange resins) of not being deformed under high pressure. Figure 1.12 is a diagrammatic representation of these materials, in which the layer thickness is roughly 1 μ.

POROUS LAYER BEADS

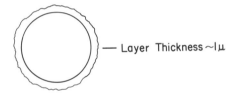

— Layer Thickness ~1μ

Figure 1.12. Diagrammatic illustration of porous layer beads. Particle diameter ~ 40 μ, porous shell ~ 1 μ.

By eliminating the deep pores, PLB diminish the path length for diffusion in the stagnant mobile phase. Diffusion can occur only in the thin layer on the surface, so that rapid mass transfer takes place. Consequently, rapid velocities can be achieved while conditions close to equilibrium are maintained. Although PLB can be used in LLC or LSC, the high activity of surface should be noted (17). Thus, if one passes a mobile phase presaturated with the stationary phase, as in LLC, over dry PLB, a steady state is reached with a constant amount of

liquid phase on the adsorbent. Consequently, there is
a limit to the liquid loadings that can be used.*

From a theoretical point of view, the thinner the
layer, the faster will be the mass transfer. Horvath
and Lipsky (28) have shown that, for constant k' and
particle diameter, the ratio of C terms for diffusion
mass transfer for PLB versus spherical beads is

$$\frac{C_{PLB}}{C_{sphere}} \cong 15\delta_F^3 (1 - 2\delta_F) \qquad (1.34)$$

where δ_F is the fractional thickness of the layer. From
this equation it is seen that very thin layers of porous
material are to be desired in terms of speed and efficiency. However, as the layer becomes thinner the
capacity of the PLB decreases, so that its separation
potential is less. Thus a compromise must be made between layer thickness and capacity (k' values). Undoubtedly, the more roughened surface incorporated into the
thin layer (i.e., the higher the surface area of the
layer) the better will be the capacity. All types can
be used to achieve separation. However, one word of
caution is required. Since the various PLB supports
differ in activity, the α values obtained in LLC at steady state will differ for each bead because mixed mechanisms of retention occur (17). Other chapters in this
book will discuss these materials in detail.

The pore depths of the support materials can also be
decreased by simply decreasing the particle diameter.
Whereas ordinary high-speed LC uses particle diameters
of 30-40 μ, Scott has successfully employed ion-exchange
beads of 5 μ diameter for various amino acid analyses
(30). The original work of Hamilton in the use of small-
particle ion-exchange beads is also noteworthy (31). In
addition, Huber (32) has employed 10 μ-diameter diatomaceous support particles. Finally, Piel in 1966 (33)

---

*One can work above the steady-state loading; below
steady state, however, the adsorbent picks up stationary
liquid.

used submicron particles for simple LC separations. The
use of particles of very small diameters creates certain
problems, however, and these will be discussed in the
next section.

2. Packing Characteristics

A central difference between liquid and gas chromatography is the major role played by band-broadening
mechanisms in the mobile phase in packed LC columns. As
a result, column-packing features are very important in
LC. In this section these characteristics will be examined as they relate to high-speed analysis.

It is generally agreed that spherical support particles are best for high-efficiency chromatography because they pack more densely and reproducibly than
irregularly shaped materials. For a regular packed column ($d_c/d_p > 10$), dense packing will result in better
radial mixing of the mobile phase via disruptions of the
flow paths by the intervening particles. In the same
regard, the particle-size distribution should be narrow.
A wide distribution will result in segregation between
large and small particles in the column. Since the permeability is proportional to d_p^2, such segregation will
create zones of different mobile-phase velocities in the
column. If the segregation occurs throughout the length
of the column, severe velocity inequalities will result.
Relaxation of the inequalities by diffusion and convection may then be difficult, especially in high-speed LC.

Because of the importance of the mobile-phase terms
in LC, the column-packing procedure has come under careful scrutiny. Snyder (34) has reviewed the various
procedures. The aim of any procedure is to create a
uniform cross-sectional density of support material and
to pack the material as densely as possible. For the
small particle diameters used in high-speed LC, there is
some difficulty in packing the support. Above 20 μ, dry-
packing the support works well, if care is taken. For
example, one should not apply a vibrator directly to a
point on the column but rather vibrate the column uniformly. Gentle tamping with a glass rod to compact the

support is also found to be useful.

Experience has shown that it is difficult to achieve high efficiency and a reproducible column by dry-packing particles smaller than 20 μ. Snyder has found that slurry packing works well in this small-particle region (34). Apparently, the surface energies of the particles become sufficiently great at these small diameters that significant agglomeration of particles occurs in dry packing (35). Wet packing provides films of liquid around the particles, thus preventing them from adhering to one another.

In high-speed LC, it has been found experimentally that column diameters of 1-3 mm are best (27). In classical LC, 8-10 mm or even larger diameters are employed. One of the reasons that the narrow columns are more efficient is that the distance is less over which radial mixing needs to occur. Relaxation of velocity inequalities can then take place more readily. In addition to the improved efficiency (at a given mobile-phase velocity), narrow columns produce a higher velocity for a given flow rate, which is, of course, desirable in high-speed work. Apparently, at diameters below 1 mm column packing becomes more difficult to achieve reproducibly. Furthermore, the mobile-phase volume, V_M, decreases substantially, the narrower the column, given a constant column length. Thus, extra-column effects play a more important role when narrow columns are used.

Knox and Parcher (20) have introduced a novel idea for LC--the infinite-diameter column. Here the sample is injected in a narrow plug in the axial center of a column. Because diffusion is slow in liquids, the band will require a certain amount of time to reach the column walls. Presumably, velocity inequalities in the center of the column will be less than those caused by wall effects. Thus enhanced efficiency ought to be achieved if the band elutes from the column before reaching the walls. Knox and Parcher found this to be the case for unretained peaks. A similar method for enhancing efficiency is to collect only the center portion of the band as it elutes from the column in ordinary LC

operation. Again the broadening effects on the band
caused by velocity inequalities will be lessened. Clearly this approach is difficult to apply in analytical-
scale high-speed LC. However, collection of only the
center portion of a band has potential preparative-scale
application (36).

As previously noted, $H \sim d_p^2$ in GC through the mobile-
phase mass transfer term. In addition, the stationary-
phase mass transfer terms can be related to d_p^2; for
example, Purnell (37) has related the film thickness
parameter, d^2, in the stationary-phase liquid diffusion
term to d_p^2. Analysis time is also related to d_p^{-2}, as
can be seen by substituting $H = Cv$ (GC) into the N_{eff}/t
equation 1.32.

$$\frac{N_{eff}}{t} = \frac{1}{C} \frac{k'^2}{(1 + k')^3} = \frac{D_m}{C'd_p^2} \frac{(k')^2}{(1 + k')^3} \qquad (1.35)$$

It would thus seem profitable in GC to make the particle
diameter as small as possible (along with decreasing the
column length). However, from a convenience point of
view, researchers have found that column lengths of
roughly 1 meter are best, so that particle sizes of
150 μ are selected.

In LC*, smaller particles (relative to those used in
GC) are selected because D_m is much lower in liquids than
in gases. In LC with porous layer beads, particle diameters of roughly 30-40 μ are selected. Although it is
desirable from a theoretical point of view to use smaller
particles and shorter column lengths, the increased difficulty in dry packing and the very important increase in

---

*Equation 1.35 is not strictly correct for LC since H
$= (1/A + 1/C_M)^{-1}$. Snyder (34) has noted empirically
that $N = \theta v^{0.4}$ in some high-speed LC systems. The general conclusion (i.e., faster analysis with smaller d_p),
however, is not changed.

extra-column effects make the 30-40 μ diameter seem at present to be a practical compromise. With ion-exchange beads, as noted, the particle diameter is 5 μ and packing is done by slurry. However, the smaller intraparticle diffusion path lengths are required for successful high-speed work.

Because of particle agglomeration and coupling of the A and C_M terms, one finds in LC that $H \sim d_p^{1.6}$ for d_p < 80 μ (16). Snyder (34) found $d_p^{0.8}$ dependence for silica; however, his particles were not spherical and hence may not have been packed densely enough. Since $K^o \sim d_p^2$, the lower dependence of H on d_p means that an increasing price is paid in pressure for the added speed attained by the smaller particle sizes (assuming L and v are constant).

The predominance of mobile-phase effects in LC means that H is independent (for the most part) of k', so that the optimum k' is roughly 2. Recently Halasz et al. (38) introduced heavily loaded LC columns in which the ratio of stationary liquid phase to solid support is 1:1. The liquid phase is oxydipropionitrile, and the support is Porasil® A (350 m^2/g). In such a column, the stationary-phase mass transfer term becomes equally important with the mobile-phase terms and a decided dependence of H on k' occurs. From the functionality in equation 1.22, $k'/(1 + k')^2$, it is seen that the C term decreases with increasing k' beyond k' = 1. It is only in such special-ly designed columns that stationary-phase terms play a role in band broadening. It should be noted that heav-ily loaded LC columns have the advantages of high k' values and less difficulty with bleeding of the liquid phase. For further details see ref. 38.

Finally, Table 1.2 compares a variety of GC and LC column types in terms of effective plates per second and effective plates per second times the column permeabil-ity. The latter term can be regarded as a performance factor for comparing column types. The sources of the data are given in the table, and the original references can be found from these sources. A word of caution is required in the interpretation of the data--namely, that

both quantities are dependent on k', among other factors.
An attempt has been made to select typical values to give
a reasonable picture; however, each value should be considered only as approximate.*

Table 1.2

Comparison of Column Types in GC and LC

Column Type	N_{eff}/t	K^o (N_{eff}/t) x 10^{+7} cm^2/sec
A. GC ([40])		
Classical packed (d_p = 130 μ)	10	2.4 x 10^{+2}
Open tubular (d_c = 0.25 mm)	25	4.7 x 10^{+3}
Packed capillary (d_p = 10 μ)	40	2.7 x 10^{+2}
Porous Layer Beads (d_p = 90 μ)	50	7.5 x 10^{+1}
Aerogel column (d_p = 1 μ)	156	1.3 x 10^{+2}
High-efficiency open tubular (d_c = 0.03 mm)	2000	7.4 x 10^{+3} ([39])
B. LC		
Classical LC (d_p = 150 μ)	0.02	4.5 x 10^{-2} ([34])
Silica (d_p = 20 μ)	2	8.0 x 10^{-2} ([34])
PLB Zipax® (d_p = 27 μ)	8	6.8 x 10^{-1} ([17])
PLB Corasil®-I (d_p = 28-37 μ)	2	3.4 x 10^{-1} ([17])

Looking first at GC columns, we see that the classical packed column gives the lowest value of N_{eff}/t. On

*K^o N_{eff}/t is only one consideration in the selection of
a column. Other factors include (1) capacity, (2) simplicity, and (3) reproducibility. However, N_{eff}/t and
K^o do give some indication of the speed characteristics
of a column and the price in pressure that must be paid
for this speed.

the basis of N_{eff}/t, (1) capillary, (2) packed capillary, and (3) PLB are similar. Aerosil® (aerogel column) is a special material of high surface area and high permeability which results in a very high N_{eff}/t. Multiplication of N_{eff}/t by K^0 changes the picture somewhat. The high permeability of the open tubular column overcomes the lower N_{eff}/t to result in the highest performance parameter. Also included in the table is the best column achieved to date in GC: an open tubular column of very narrow diameter, made by Desty et al. (39). Over 2000 effective plates per second were achieved. Unfortunately the column is most difficult to make and is included in the table only for academic interest.

In regard to LC, the very poor quality of classical LC columns is clearly seen. Only 0.02 effective plate per second can be generated. Almost 7 hrs would be needed to achieve 500 effective plates (a reasonable number for general separation problems). Changing to smaller particle sizes (e.g., 20-μ silica beads) results in a hundred-fold increase in effective plates per second but only a two-fold increase in effective plates per second times column permeability. Increased speed of analysis is bought at the price of pressure. Finally, recent results obtained with commercially available PLB supports, Zipax® and Corasil® I, are shown. The performance of these PLB (in terms of $N_{eff}/t \times K^0$) is significantly better than results obtained with classical LC and silica. If viscosity differences are taken into account, PLB of LC are approximately equal to PLB of GC in terms of $K^0(N_{eff}/t)$. However, a factor of 5-10 greater effective plates per second is achieved when PLB is used in GC. Nevertheless, it can be seen that the PLB of LC are close to classical packed GC columns in terms of N_{eff}/t. Thus a great deal of progress has been achieved in high-speed LC.

In the future, it is expected that with new column designs (smaller particle sizes, irregular packed columns)

N_{eff}/t will increase in LC. The day is not far away when high-speed LC will compare favorably with GC in terms of speed of analysis.*

CONCLUSION

Although it is clear from the discussion in this paper that high-speed LC has progressed quite far, it is equally obvious that new developments should rapidly appear in the next few years. Several areas where these developments are likely to occur can be mentioned.

First, as noted, one of the problems in high-speed LC is the low permeability of the columns, arising from the small particle diameters and high mobile-phase viscosity. Use of irregular packed columns ($d_c/d_p < 5$) would increase the permeability by at least an order of magnitude, while concomitantly increasing efficiency. If irregular packed columns can be reproducibly made and if they can maintain their characteristics with high pressure and flow rate, such columns might be preferred when a large number of effective plates are required (e.g., in a difficult separation or multicomponent analysis).

Second, although column characteristics have been emphasized in this chapter, it is clear that theory can be very helpful in the selection of mobile and stationary phases for the achievement of given separations. At present, the ability to predict appropriate phases is relatively poor. Simple models are needed that will permit the use of property data for pure compounds and the selection of the best system for a given separation. A simple and workable solvent classification scheme would be most useful to practicing chromatographers.

---

*This discussion has of course assumed a constant α for both GC and LC. Although it is difficult to compare the two methods, LC does have the greater potential for selectivity through the use of the mobile phase. Thus, in many cases, an equivalent number of effective plates per second for a GC and a LC column may mean a faster separation with the latter because of larger α value.

Third, it is clear that the developments in analyti-
cal-scale LC will influence preparative-scale LC. Indeed
there are indications that the loss in efficiency in
using wide-diameter columns in high-speed LC is not as
great as originally thought. It is certain that improvements in preparative LC will have a major impact on
laboratory research, since classical LC is already used
in this application.

Fourth, high-speed GPC is an area of obvious development. There have already been some efforts in this
direction (41). Indeed, it is clear that rapid molecular weight analysis of polymers is possible.

Fifth, developments can be expected in detectors.
These devices have not been discussed in this chapter;
however, it is clear that they play a central role in
determining the ability of a system to carry out a separation by LC. This important topic is discussed in
detail in Chapter 3.

Finally, for those desiring a more detailed account
of theory in LC, ref. 42 cites several sources of information.

REFERENCES

1. E. Heftmann, ed., "Chromatography", Reinhold, New
    York, 1967.
2. B. Karger, K. Conroe, and H. Engelhardt, J. Chro-
    matog. Sci., 8, 242 (1970).
3. E. Cremer, Th. Kraus, and H. Nau, Z. Anal. Chem.,
    245, 37 (1969).
4. J. H. Purnell, J. Chem. Soc., 1960, 1268.
5. I. Halász and E. Heine, in "Progress in Gas Chroma-
    tography," J. H. Purnell, ed., Wiley-Interscience,
    New York, 1968, p. 153.
6. A. Einstein, Z. Elektrochem., 14, 1908 (1908).
7. J. C. Giddings, "Dynamics of Chromatography," Marcel
    Dekker, New York, 1965, pp. 190-193.
8. Ibid., p. 29
9. Ibid., p. 52.
10. J. H. Knox and M. Saleem, J. Chromatog. Sci., 7,
    745 (1969); see references therein to other papers.

11. R. P. W. Scott, D. W. J. Blackburn, and T. Wilkins, J. Gas. Chromatog., 5, 183 (1967).

12. G. Deininger and I. Halász, in "Advances in Chromatography, Sixth International Symposium," A. Zlatkis, ed., Preston Technical Abstracts Co., Chicago, 1970, p. 336.

13. R. D. Conlon, Anal. Chem., 41, (4) 107A (1969).

14. J. C. Sternberg, in "Advances in Chromatography", Vol. 2, J. C. Giddings and R. A. Keller, eds., Marcel Dekker, New York, 1966, p. 205.

15. C. Horvath and S. R. Lipsky, J. Chromatog. Sci., 7, 110 (1969).

16. A. Kroneisen, Ph.D. Thesis, University of Frankfurt, 1969.

17. B. L. Karger, H. Engelhardt, K. Conroe, and I. Halász, paper presented at the Eighth International Symposium on Chromatography, Dublin, September, 1970.

18. P. C. Carman, "Flow of Gases Through Porous Media," Butterworths, London, 1956, p. 8.

19. I. Halász and P. Walkling, J. Chromatog. Sci., 7, 129 (1969).

20. J. H. Knoz and J. F. Parcher, Anal. Chem., 41, 1599 (1969).

21. J. H. Knox and M. Saleem, J. Chromatog. Sci., 7, 614 (1969).

22. J. C. Giddings, Anal. Chem., 39, 1027 (1967).

23. C. G. Horvath and S. R. Lipsky, Anal. Chem., 39, 1893 (1967).

24. M. J. E. Golay, in "Gas Chromatography, 1960," R. P. W. Scott, ed., Butterworths, London, 1960, p. 139.

25. J. Bohemen and J. H. Purnell, J. Chem. Soc., 1961, 360.

26. I. Halász and C. Horvath, Anal. Chem., 36, 1178 (1964); I. Halász and C. Horvath, Anal. Chem., 36, 2226 (1964).

27. J. J. Kirkland, J. Chromatog. Sci., $\underline{7}$, 7 (1969);
     J. J. Kirkland, J. Chromatog. Sci., $\underline{7}$, 361 (1969);
     J. J. Kirkland and J. J. DeStefano, J. Chromatog.
     Sci., $\underline{8}$, 309 (1970).
28. C. Horvath, B. Preiss, and S. R. Lipsky, Anal. Chem.,
     $\underline{39}$, 1422 (1967); C. Horvath and S. R. Lipsky, Anal.
     Chem., $\underline{41}$, 1227 (1969).
29. J. N. Little, D. F. Horgan, and K. J. Bombaugh,
     paper accepted by J. Chromatog. Sci.; Waters
     Associates, "Chromatography Notes," $\underline{1}$, 1 (1970).
30. C. D. Scott, J. E. Attril, and N. G. Anderson,
     Proc. Soc. Exptl. Biol. Med., $\underline{125}$, 181 (1967).
31. P. B. Hamilton, in "Advances in Chromatography,"
     Vol. 2, J. C. Giddings and R. A. Keller, eds.,
     Marcel Dekker, New York, 1966.
32. J. F. K. Huber, prviate communication.
33. E. V. Piel, Anal. Chem., $\underline{38}$, 670 (1966).
34. L. R. Snyder, J. Chromatog. Sci., $\underline{7}$, 352 (1969).
35. I. Halász, private communication.
36. J. J. DeStefano and H. C. Beachell, J. Chromatog.
     Sci., $\underline{8}$, 434 (1970).
37. J. H. Purnell, "Gas Chromatography," John Wiley,
     1962, p. 128.
38. I. Halász, H. Engelhardt, J. Asshauer, and B. L.
     Karger, paper submitted to Anal. Chem.
39. D. H. Desty, A. Goldup, and W. T. Swanton, in "Gas
     Chromatography", N. Brenner, J. E. Callen, and
     M. D. Weiss, eds., Academic, New York, 1963, p. 105.
40. G. Guiochon, in "Advances in Chromatography", Vol.
     8, J. C. Giddings and R. A. Keller, eds., Marcel
     Dekker, New York, 1969.
41. J. N. Little, J. L. Waters, K. J. Bombaugh, and
     W. Pauplis, J. Polymer Sci., $\underline{7}$, 1775 (1969).
42. (i)   L. R. Snyder, "Principles of Adsorption Chro-
            matography," Marcel Dekker, New York, 1965.
     (ii)   L. R. Snyder, Anal. Chem., $\underline{39}$, 698, 705 (1967).
     (iii) J. F. K. Huber, J. Chromatog. Sci., $\underline{7}$, 85
            (1969).
     (iv)   J. F. K. Huber and J. Hulsman, Anal. Chim.
            Acta, $\underline{38}$, 305 (1967).

(v)   J. C. Giddings, "Dynamics of Chromatography",
        Marcel Dekker, New York, 1965.

(vi)   T. W. Smuts, F. A. van Niekerk, and V.
        Pretorius, J. Gas Chromatog., $\underline{5}$, 190 (1967).

(vii) References $\underline{10}$ and $\underline{15}$.

43. J. H. Knox and L. McLaren, Anal. Chem., $\underline{35}$, 449
     (1963).

CHAPTER 2

Apparatus for High-Speed Liquid Chromatography

Richard A. Henry

A. INTRODUCTION

A primary consideration when constructing or buying a liquid chromatographic system is the type of liquid chromatography to be done. It is difficult to design a system that will be ideal for steric exclusion, large-scale preparative, and high-speed analytical chromatography. The purpose of this chapter is to acquaint the reader with the apparatus that is currently available and to provide him with insight into the advantages and disadvantages of various designs so that he can build or buy the liquid chromatograph that best suits his needs.

A block diagram of a typical liquid chromatographic system is shown in Figure 2.1. The hardware that makes up each block will be discussed in the order in which it appears in the diagram. Detectors are treated separately in detail in Chapter 3.

B. SOLVENT RESERVOIRS

Several factors are important when selecting a solvent reservoir for a liquid chromatographic system. The first consideration should be the volume required. This will vary greatly, of course, with the particular type of liquid chromatography to be practiced. For example, small reservoirs holding approximately 1 liter are optimum for high-speed, analytical liquid chromatography in which flow rates are typically 1-2 ml/min and separations are completed in a few minutes. Larger-volume reservoirs should be selected for chromatographic systems which require high flow rates, such as those used in preparative chromatography with large-diameter columns.

Another consideration which may be important is the number of separate flow systems to be fed from the reservoir. For example, there may be occasion to operate a reference column along with the analytical column, and

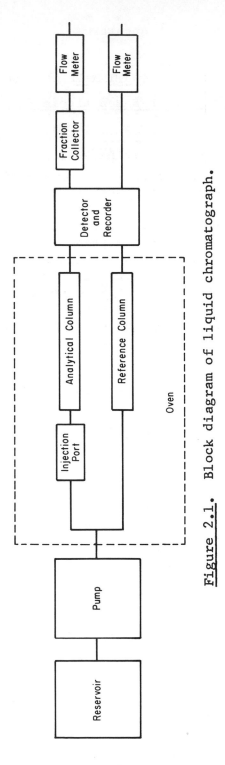

Figure 2.1. Block diagram of liquid chromatograph.

56

this will effectively cut in half the useful capacity of the reservoir for analytical work unless provision is made to return the mobile phase to the reservoir.

1. Reservoir Materials

Type 304 or 316 stainless steel makes an excellent reservoir material for liquid chromatography because it is inert to almost any mobile phase and is not subject to breakage. Resistance to breakage is a desirable safety advantage over glass when flammable or toxic solvents are being used as the mobile phase. Also, the mobile phase can be safely degassed by heating and/or applying vacuum directly to the reservoir. A reservoir or series of reservoirs constructed of stainless steel can be designed using commercially available valves and fittings, so that only stainless steel and Teflon® come into contact with the mobile phase. Figure 2.2 shows a typical stainless steel reservoir. Under proper conditions glass or an inert polymer such as polytetrafluoroethylene can also be used as the reservoir construction material.

2. Reservoir Options

Options which may be included in the reservoir are a heater, a stirrer, and inlets for vacuum and nitrogen purge. Perhaps the most common operation that can be carried out in the reservoir is degassing the mobile phase. Degassing is important with water and other polar solvents to keep bubbles from forming in the detector; bubbles can be especially troublesome with low-dead-volume detectors. Degassing to prevent the formation of bubbles can be accomplished by pulling a vacuum on the filled reservoir for a few minutes while stirring the mobile phase vigorously. Complete removal of dissolved oxygen from some solvents may require heating and nitrogen purge. Degradation of easily oxidized samples or stationary phases can be prevented by removal of the last traces of oxygen from the mobile phase with a nitrogen purge. A nitrogen blanket on the reservoir will

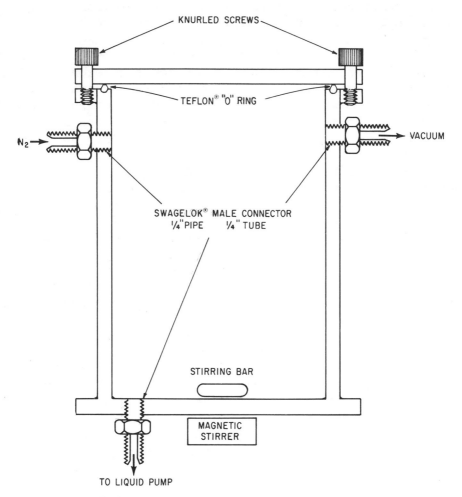

Figure 2.2. Stainless steel reservoir for liquid chromatography.

keep more soluble gases, such as oxygen, from redissolving in the mobile phase after degassing has been completed, and will also act as a safety precaution when the reservoir is being heated.

It should be pointed out that air can leak in at a column fitting without mobile phase leaking out. When

bubbles appear in the detector and no leak is visible in
the system, fittings should be tightened, especially at
the column outlet and detector inlet.

Other factors may become important when the reservoir functions as more than just a solvent container.
For example, the reservoir can also serve as a liquid
pump if it is constructed of compressible plastic or
steel bellows and is placed in a pressurized air chamber (1).

C. PUMPING SYSTEMS

Several factors must be considered when selecting a
pumping system for liquid chromatography. Perhaps the
most important is the maximum column inlet pressure required. Although many excellent liquid chromatographic
separations have been carried out at inlet pressures
below 1000 psig, the small particle sizes and small-
diameter columns currently being used for rapid, highly
efficient separations may require inlet pressures of up
to 5000 psig.

Other important considerations may be pulsating or
pulse-free operation, constancy of flow, and maximum
deliverable flow. To be useful for liquid chromatography,
a pump should be pulse-free or have a pulse-dampening
device, be able to deliver mobile phase with a constancy
of at least 2-3%, and have a flow delivery range of at
least 0-10 ml/min. The pump and other parts of the system must also be able to resist the solvents used as
mobile phases. Pumps which meet these requirements can
be grouped into two major categories: mechanical and
pneumatic.

1. Mechanical Pumps

One of the most common types of mechanical pumps is
the screw-driven syringe type, which has been incorporated into at least two commercial instruments for high-
speed liquid chromatography (Nester-Faust, Newark, Del.;
Varian Aerograph, Walnut Creek, Calif.). The advantages
of this type of pump are that it can deliver a pulse-free

supply of liquid at a constant rate and, with metal construction, can withstand very high column inlet pressures. A disadvantage is that it has limited solvent
capacity and must be stopped for refilling and resetting.
This difficulty can be avoided by using two pumps alter-
nately so that one fills while the other empties.

Other mechanical pumps in use are the reciprocating
piston (Milton-Roy Co., St. Petersburg, Fla.) and the
diaphragm (Whitey Research Tool Co., Emeryville, Calif.;
Orlita Dosiertechnik KG, Giessen, Germany) types, which
deliver a pulsating supply of liquid. Diagrams of two
commercial reciprocating pumps are shown in Figures 2.3
and 2.4. These pumps supply a pulsating mobile phase
which causes periodic noise especially disturbing to the
bulk property detectors. To reduce this noise, a dampening device is used to eliminate pulsations. One arrangement consists of 10-20 ft of 1/16-in.-o.d., 0.010 to

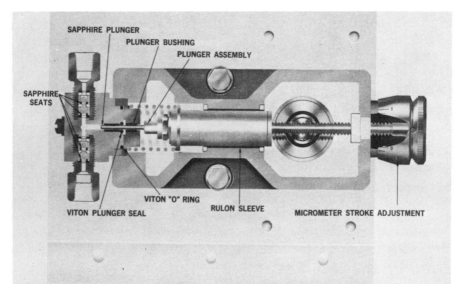

Figure 2.3. Reciprocating piston pump. (Courtesy
of Milton-Roy Co.)

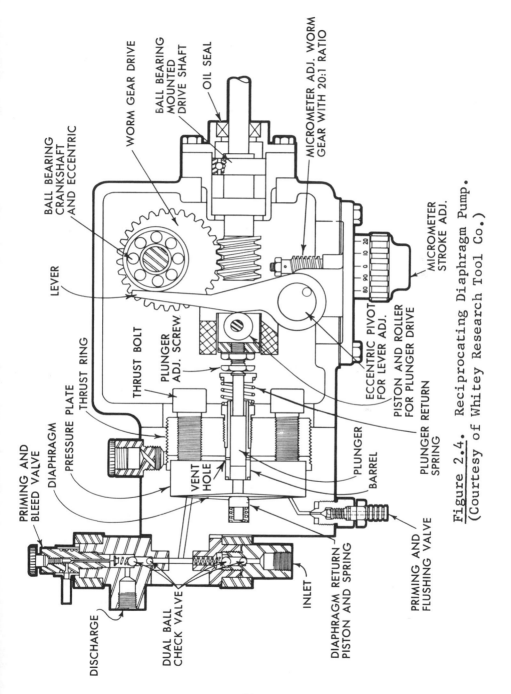

PRIMING AND
BLEED VALVE

DIAPHRAGM

PRESSURE PLATE
THRUST RING

DISCHARGE

DUAL BALL
CHECK VALVE

INLET

DIAPHRAGM RETURN
PISTON AND SPRING

PRIMING AND
FLUSHING VALVE

VENT
HOLE

THRUST BOLT

PLUNGER
ADJ. SCREW

LEVER

BALL BEARING
CRANKSHAFT
AND ECCENTRIC

WORM GEAR DRIVE

BALL BEARING
MOUNTED
DRIVE SHAFT

OIL SEAL

MICROMETER ADJ. WORM
GEAR WITH 20:1 RATIO

MICROMETER
STROKE ADJ.

ECCENTRIC PIVOT
FOR LEVER ADJ.

PISTON AND ROLLER
FOR PLUNGER DRIVE

PLUNGER RETURN
SPRING

PLUNGER

BARREL

Figure 2.4. Reciprocating Diaphragm Pump.
(Courtesy of Whitey Research Tool Co.)

61

0.020-in.-i.d. stainless steel tubing wound into a 4-cm-diameter coil. The required dimensions of the coil will vary with the amplitude and frequency of the pulsations. Since the pulses are absorbed by the flexing of the coil, it is important to allow the coil to hang free. Most types of dampening devices increase the total volume of the system and, therefore, increase the time required for complete mobile-phase changeover.

A specific advantage of the reciprocating pump is that its internal volume is small and its delivery is continuous, so that there is virtually no restriction on the size of the reservoir. A low-dead-volume reciprocating pump also has an advantage in steric exclusion chromatography in that samples can be recycled back to the column with only a small loss in efficiency. (Sample recycling in steric exclusion chromatography is discussed in Chapter 7.) Several commercial instruments utilize this pump design (Waters Associates, Framingham, Mass.; Varian Aerograph, Walnut Creek, Calif.; Siemens Aktiengesellschaft, D7500 Karlsruhe 21, P.O.B. 211080, Federal Republic of Germany).

An advantage of both the reciprocating and the screw-type mechanical pumps is that they deliver constant volume regardless of small variations in the pressure drop across the rest of the system. With these constant-volume mechanical pumps, a relief valve or rupture disc must be included to protect components from damage if a blockage occurs in the system.

2. Pneumatic Pumps

Pneumatic pumps are operated by gas pressure working on a suitable collapsible container or piston that pressurizes the mobile phase. Jentoft and Gouw (2) have described a pneumatic pump which utilizes mercury as a piston to transfer pressure from a gas cylinder to the mobile phase. In the simplest form of this pump, a plastic bottle or stainless steel or Teflon® bellows is placed inside an airtight container and gas pressure is applied to the inside of the gas container and the outside of the liquid container (1). The maximum pressure

that can be delivered by this type of pump is limited by
the strength of the container and the pressure of the gas
supply, which, for normal tanks, is about 3000 psig.
This type of pneumatic pump is used in at least one commercial instrument (Pye Unicam, Cambridge, England).

Most commercial gas-driven piston pumps amplify the
gas pressure by using a large-area pneumatic piston and
a small-area liquid piston. A diagram of a commercially
available pneumatic amplifier pump is shown in Figure
2.5. An attractive feature of the pneumatic pump illustrated is that it has a power return stroke that rapidly

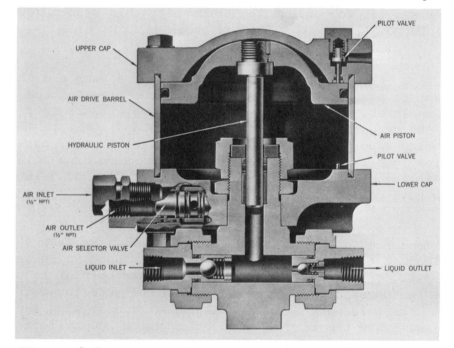

Figure 2.5. Pneumatic amplifier pump. (Courtesy of
Haskell Engineering and Supply Co.)

refills the liquid chamber from an external reservoir,
so that the base line is interrupted only momentarily
when the pump refills (3). One commercial instrument
(E. I. du Pont de Nemours & Co., Instrument Products
Div., Wilmington, Del.) employs a pneumatic amplifier

pump that develops about a 50 to 1 compression ratio, so
that 50 psig gas inlet pressure yields 2500 psig liquid
outlet pressure.

Advantages of pneumatic pumps are that high pressures
can be easily achieved and flows are pulse-free and relatively constant. Another advantage is that flow can be
easily programmed by varying the gas supply with a pressure programmer similar to that used in gas chromatography (4,5). The disadvantages are that the pumps can
deliver only a limited volume of liquid before refilling
and that the flow rate remains constant only as long as
the pressure drop across the chromatographic system does
not change.

3. Line Filters

Line filters should be used between the pump and the
column to prevent particulate matter from clogging the
column inlet. Filters are commercially available in a
variety of pore sizes (Nupro Co., Cleveland, Ohio). The
porosity should be small enough to collect harmful particles, yet large enough so that the pressure drop across
the filter element is negligible. A pore size of 5-10 μ
is a good compromise.

4. Back-Pressure Devices

In a flowing system, the pressure will vary from a
maximum at the column inlet to nearly atmospheric in the
detector cell at the column outlet. It is helpful to be
able to increase the back pressure on the detector cell
to at least 20 psig to prevent bubbles from forming and
to facilitate the removal of bubbles which become lodged
in small-volume detector cells. One way of applying and
monitoring this back pressure is by installing a meter-
ing valve (Whitey Research Tool Co., Emeryville, Calif.;
Nupro Co., Cleveland, Ohio) and pressure gauge downstream from the detector cell. With this arrangement,
the desired back pressure can be selected by partially
closing the valve and observing the gauge. A second
method uses a downstream check valve (Nupro Co.,
Cleveland, Ohio) to apply a constant back pressure to

the cell. Check valves should be chosen so that their internal parts are resistant to the solvents which will be used as mobile phase.

D.   SAMPLE INTRODUCTION DEVICES

There are two basic methods of introducing a sample to the chromatographic column: injection ports and sample valves.

1.   Injection Ports

Injection ports can, in turn, be grouped into two major types: those in which the sample is swept into the column by the mobile phase, and those which allow the sample to be deposited directly in the packing.

On-Column Injection Ports. The highest separation efficiency can often be obtained by injecting the sample directly into the center of the column packing. A diagram of a simple on-column injection port is shown in Figure 2.6. In this arrangement, the syringe needle extends through the septum and into the column packing. Large or small samples can be deposited, the size of the sample being limited primarily by the capacity of the packing and the diameter of the column. A troublesome feature of on-column injection is that syringes tend to plug frequently because the diameter of the column-packing material is small enough to be forced into the syringe needle. (This can be avoided by designing the injection port so that the syringe needle stops in a glass wool plug just before the column packing.) Also, septum material can work into the column inlet and disturb the packed bed. On-column injection is discussed further in Chapter 5.

Swept Injection Ports. In a swept injection port, the sample is deposited just before the column inlet and swept into the packing by the mobile phase. It is important to minimize band spreading before the sample reaches the column packing; consequently, the injection port which allows the sample to be swept into the column must be carefully designed to minimize dead volume. For

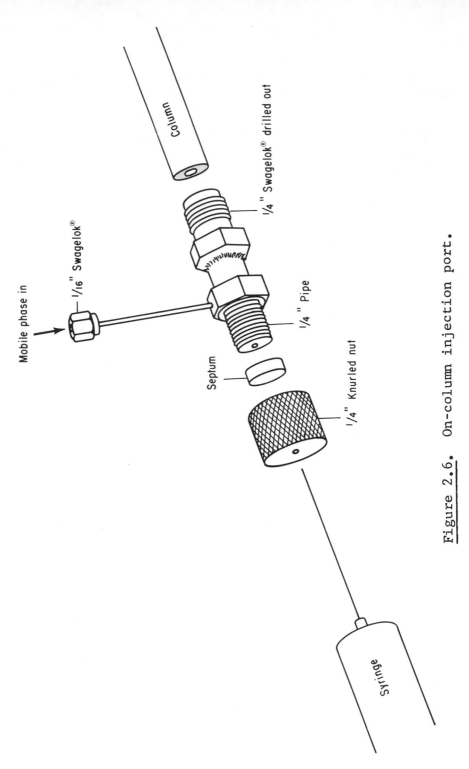

Figure 2.6. On-column injection port.

this reason, it is difficult to design an injection port that will efficiently sweep into the column both the small samples of a few microliters that are typical in most analytical work and the large samples of 50 or 100 μl that may be desirable for small-scale preparative work or trace analysis. For either small or large samples (but not both) a swept injection port can closely approach the efficiency of on-column injection. In fact, for compounds which have k' > 1, the efficiency of a well-swept injection port is virtually identical to that of on-column injection. For unretained and slightly re-tained compounds, k' < 1, on-column injection can result in slightly higher efficiency (6,7).

An advantage of using an instrument with the swept injection port design is that the column packing can be firmly captured between two porous plugs. In this case there is little likelihood of the packed bed coming loose during column shipment or storage, and no septum material or other particulate matter can work its way into the packed bed.

Stop-flow or on-stream injection can be carried out with either type of injection port. Stop-flow injection with a low-pressure syringe can be used with good re-sults when a high-pressure syringe or sample valve is not available. A stop-flow technique may also be desir-able when very high inlet pressures (above 2000 psig) are being used and when large samples are being intro-duced to the column. The efficiency of stop-flow injec-tion is virtually identical to that of on-stream injection because of the slow rate of sample diffusion in a liquid mobile phase.

2. Sample Valves

Another method of sample introduction utilizes a valve arrangement. Sample valves are often used in steric exclusion chromatography, where large sample sizes are common, and in very-high-pressure liquid chro-matography (above 2000 psig), where sample injection with a syringe becomes difficult. Sample valves are

also employed when high repeatability of sample introduction is required. A six-port injection valve that can
be used at inlet pressures up to 1500 psig is shown in
Figure 2.7. In one handle position the mobile phase

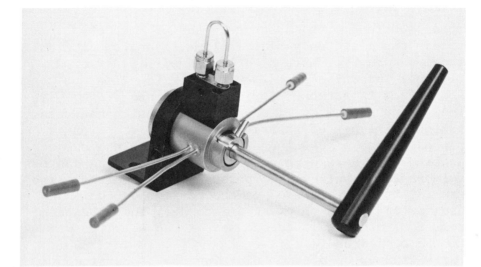

Figure 2.7. Six-port sample valve. (Courtesy of
Valco, Inc.)

flows directly through the valve and into the chromatographic column, while in the other handle position the
mobile phase sweeps the sample from the sample loop
into the column. A six-port valve designed for operation at pressures up to 5000 psig has been recently
described in the literature (8). When a large-volume
loop is used with the six-port valve, various sample
sizes can be introduced by opening the valve only long
enough to sweep part of the coil volume into the column.
If flow rates are constant, very reproducible samples
can be introduced by this method. A simple four-port
valve for the introduction of sample sizes of a few
microliters at inlet pressures up to 2500 psig is also
available (Valco, Inc., Houston, Tex.). A sample valve
can be designed with minimum dead volume by extending
the valve outlet tubing into the column packing.

An advantage of sample valves is that they are easily
adapted to automatic repetitive analysis. Although automatic sampling systems for steric exclusion chromatography have been available for some time, sampling systems
for high-speed liquid chromatography in which the sample
valve is operated mechanically (Valco, Inc., Houston,
Tex.) or pneumatically (Hamilton Co., Whittier, Calif.)
are just now coming on the market.

E. COLUMNS

1. Column Materials

The type of tubing used to construct high-performance
liquid chromatographic columns greatly affects their
efficiency (9). Figure 2.8 compares the plate heights
obtained from columns made of Trubore® glass, stainless
steel, and precision-bore stainless steel tubing. These

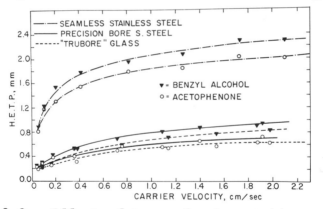

Figure 2.8. Effect of type of column tubing on column efficiency. Columns: Seamless stainless steel,
500 mm x 3 mm i.d. x 6.35 mm o.d.; precision-bore
stainless steel, 500 mm x 3.2 mm i.d. x 6.35 mm
o.d.;glass, 500 mm x 3.0 mm i.d. x 10 mm o.d.
Packing: 1% β,β'-oxydipropionitrile on 20-37 μ CSP
support; carrier: hexane; sample: 3 μl of 1 mg/ml
acetophenone, 5 mg/ml benzyl alcohol in hexane;
detector: 0.2 absorbance, full scale; temperature:
27°C. (Reproduced from ref. 9 by permission of the
publisher.)

data show that columns made from the Trubore® glass and
precision-bore stainless steel are significantly more
efficient than those made with ordinary stainless steel
tubing using conventional packing techniques. Similar
trends have also been demonstrated with columns packed
with diatomaceous earth supports (7). Apparently, the
very smooth walls of the Trubore® glass and the precision-bore stainless steel tubings result in a more homogeneous dry-packed column. This belief is substantiated
by the fact that ordinary stainless steel columns having
a very smooth internal coating of Teflon® polytetrafluoroethylene demonstrate efficiencies essentially
identical to those of precision-bore stainless steel
columns (10).

2. Column Plugs

Before packing, the column outlet should be filled
with a plug of porous metal (Mott Metallurgical Corp.,
Farmington, Conn.) or porous Teflon® (Fluoro-Plastics,
Inc., Philadelphia, Pa.). The porosity of the plug
should be just low enough to prevent the smallest packing particles from escaping from the column or plugging
the pores. Plugs should fit snugly into the column so
that they will withstand the pressures to which they are
subjected. After a column has been packed, the inlet
should also be plugged with porous metal or other suitable material to capture the packing securely in the
column. Teflon® or other soft porous plugs can be used
only at relatively low inlet pressures (below 1000 psig)
because the pores collapse at high pressures. Silanized
glass or quartz wool is preferred when an on-column
injection method is employed.

3. Column Fittings

Columns should be connected to injection ports, detectors, or other columns with fittings designed to introduce a minimum of dead volume to the system. Standard
compression fittings, such as Swagelok®, are usually
employed to seal high-pressure liquid. Fittings should

be drilled out so that the column plug is snug against
the inlet or outlet tubing, which, in turn, is flush
with the inside wall of the fitting to minimize dead
volume. A diagram of a low-dead-volume fitting is shown
in Figure 2.9.

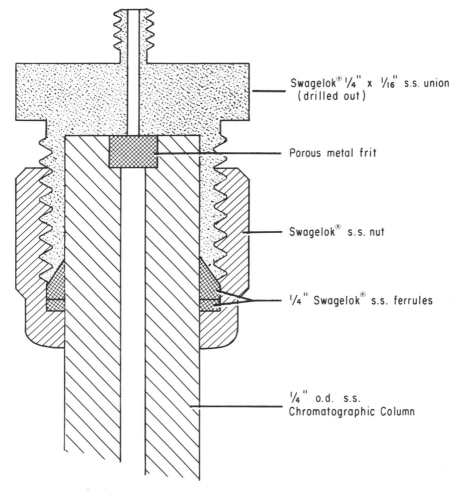

Swagelok® ¼" x ¹⁄₁₆" s.s. union
(drilled out)

Porous metal frit

Swagelok® s.s. nut

¼" Swagelok® s.s. ferrules

¼" o.d. s.s.
Chromatographic Column

Figure 2.9. Low-dead-volume column fitting.

Stainless steel ferrules should be used in all high-
pressure fittings; it should be noted that these ferrules
will swage down on steel tubing and may be difficult to

remove. Teflon® ferrules can be used in low-pressure
fittings and can be easily removed from tubing. Brass
ferrules can be substituted for steel or Teflon® in certain applications, and are easier to remove from tubing
than steel. A disadvantage of brass is that it is not
as chemically inert as stainless steel.

4. Column Configuration

Straight columns appear to be preferred for high-
performance liquid chromatography. Lengths of 50 or 100
cm are convenient and are easy to empty and repack. Columns filled with packings and then coiled have shown a
significant decrease in efficiency (9,12). Coiled columns are difficult, if not impossible, to refill by dry-
packing techniques. If filled columns must be bent to
fit a particular instrumental configuration, it appears
that they should be bent sharply at right angles to prevent a significant loss in efficieney (11).

To produce a system with a large number of theoretical plates, one may connect individual straight columns
in series to produce longer columns with little loss in
total theoretical efficiency. The requirement for this
operation is that the columns to be connected be
"matched"; that is, they should be of the same internal
diameter, particle size diameter, and efficiency (12).
Figure 2.10 illustrates the small loss in total efficiency obtained by connecting two or three columns in
series. A low-dead-volume fitting for connecting columns
in series is shown in Figure 2.11. The fitting is designed so that the connecting tubing protrudes into the
column packing. If frits are used in the column inlets,
the connecting tubing can be cut flush with the top of
the porous plug with little or no loss in efficiency.

5. Column-Packing Devices

Dry Packing. Materials which can be easily dry-packed
into columns include Zipax®* (Du Pont, Instruments Div.,

---

*Du Pont trademark.

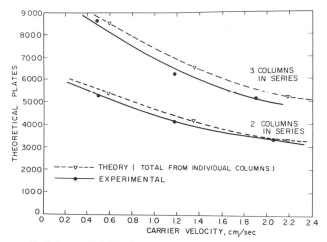

Figure 2.10. Efficiency of columns connected in series. Columns: 2.1 mm i.d., precision-bore stainless steel; packing: 0.5% β,β'-oxydipropionitrile on 20-37 μ CSP support; carrier: hexane; sample: 5 μl of 1 mg/ml benzyl alcohol in hexane; detector: 0.1 absorbance, full-scale; temperature: 25°C. (Reference as in Figure 2.11.)

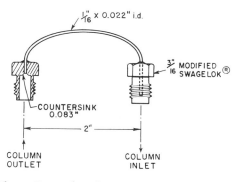

Figure 2.11. Low-dead-volume column connector. (Reproduced from ref. 9 by permission of the publisher.)

Wilmington, Del.) and Corasil®* (Waters Associates,
Framingham, Mass.) general chromatographic supports,
Pellex®** ion-exchange resins (Northgate Laboratories,
Hamden, Conn.), and porous glass (Corning Glass, Corning,
N.Y.) for steric exclusion chromatography. Although
excellent columns can be dry-packed manually, the pro-
cess is time-consuming and, in inexperienced hands, is
subject to many uncontrolled variables.

A diagram of a versatile, automatic packing device for
single columns is shown in Figure 2.12. Enough packing
to fill the column is placed in the feeding device and
is slowly fed to the column while the arm to which the
column is fastened is lifted vertically by rods attached
to a rotating motor shaft and allowed to drop on the
metal stop. The feeding device should be a container
with a restricted orifice at the bottom so that the packing is delivered slowly. Provision is usually made to
stir or vibrate the packing in the feeder to prevent
clogging of the orifice. When the column has been
filled, the inlet is plugged to capture the packing and
a fitting is added to make the column compatible with an
injection port or sample valve. The column-packing device can be designed so that both the frequency and the
amplitude of rapping can be varied to optimize column
efficiency. A plumb bob can be used to align the column
vertically, and the entire system can be enclosed and
insulated for quiet operation.

Slurry Packing. Packing materials which cannot be
dry-packed into columns must be slurry-packed. As discussed in Chapter 7, packings for steric exclusion chromatography, such as Styragel® (Waters Associates,
Framingham, Mass.) or BioBeads® (Bio-Rad Laboratories,
Richmond, Calif.), must be slurry-packed because they
swell when solvated. Ion-exchange resins, which are discussed in Chapter 8, also must be slurry-packed because
of their swelling characteristics. Even the highly

---

* Waters trademark.
**Northgate trademark.

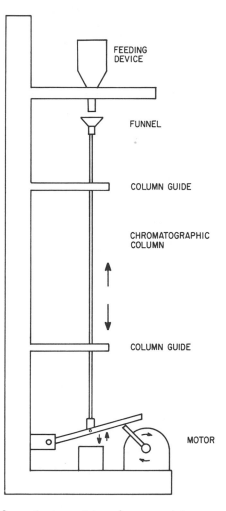

<u>Figure 2.12.</u>   Automatic dry-packing apparatus.

cross-linked resins, which have low swelling characteristics, must be slurry-packed because they are not available in a free-flowing form as are the solid-core packings. The packing of materials with chemically bonded stationary phases by a balanced-density slurry technique has also shown advantages (<u>13</u>) (see Chapter 5).

Detailed methods for packing columns for steric exclusion chromatography (<u>14</u>) and ion-exchange chromatography
(<u>15</u>) have been published. A general apparatus for slurry
packing is shown in Figure 2.13.

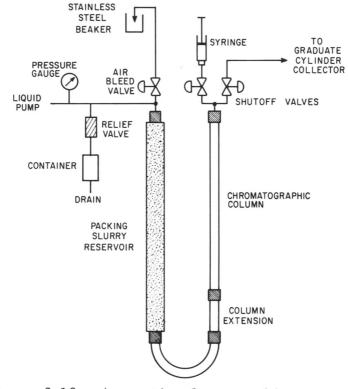

<u>Figure 2.13</u>. Automatic slurry-packing apparatus.

F. <u>COLUMN OVENS</u>

A column oven should have adequate capacity for any
column length or diameter that will be required, and
should also be large enough to accommodate more than one
column when a reference column or series of columns must
be used. When designing an oven for liquid chromatography, it is important to realize that both the heat
capacity of the circulating medium and the rate at which
it is circulated are important in determining how closely

temperature can be controlled. Air has a low heat capacity but can be circulated very rapidly, so that in many cases an air oven can do a better job of temperature control with fewer problems than a liquid bath.

1. Circulating Air Bath

Basically, the problem of temperature control is the same in liquid chromatography as in gas chromatography. The reasons for desiring close temperature control are discussed in other chapters. Although the heat capacity of a liquid mobile phase is much greater than that of a gas mobile phase, there is very little liquid mobile phase in a high-efficiency liquid chromatographic column, and the heat capacity of the column and packing will dominate the system as it does in gas chromatography. For this reason, circulating high-velocity air ovens of the type employed in gas chromatography can also be used in liquid chromatography. Oven designs and proportional controllers which will control temperatures in the region of the columns to $\pm 1\%$ are readily available. Temperature fluctuations will be considerably less than 1% inside the stainless steel tubing of the column. If the mobile-phase flow rate is high, as in the case of large-diameter preparative columns, or if the operating temperature is high, an equilibration coil consisting of about 15 ft of 1/16-in. stainless steel tubing may be placed in the oven to bring the mobile phase up to temperature before it reaches the column.

Circulating air baths have an advantage over water baths in that they respond much more quickly when the chromatographer wishes to change operating temperature. Also, air baths are more convenient when columns must be changed or leaks repaired.

When flammable solvents such as hexane are used, circulating air ovens should be purged with nitrogen until the oxygen concentration falls below 10% before operating heaters with exposed heating elements.

2. Circulating Water Bath

The only significant advantage of a water bath for
temperature control in liquid chromatography is that the
equipment is readily available in many laboratories.
Temperature control can be achieved simply by jacketing
the column and circulating liquid from a thermostated
bath. Scott (16) has described a water bath for temperature programming in liquid chromatography.

G. FRACTION COLLECTORS

1. Manual Collection

In high-speed liquid chromatography, a separation can
usually be accomplished in a few minutes with a minimum
of band spreading. For this reason, it is convenient to
collect peaks manually as they are indicated on the recorder chart. A simple three-way valve can be attached
to the detector outlet to divert the mobile phase to a
collection vessel. The fraction collection valve should
be well swept and have a small internal volume. Tubing
connecting the valve to the detector should have a small
internal diameter and be as short as possible, so that
the compound appears at the fraction collector within a
few seconds after it is seen by the detector.

2. Automatic Collection

Automatic continuous fraction collectors are used at
the outlet of liquid chromatographic columns to collect
the eluent fractions in preparative work or for additional analytical investigation. A good fraction collector is an integral part of a detection procedure when
no continuous detector is available and some property of
the collected fraction (absorbance, radioactivity, etc.)
must be measured as a function of time or mobile-phase
volume. However, the need for fraction collectors has
been reduced by the development of sensitive detectors
that continuously monitor the eluent stream and automatically indicate when compounds are emerging from the
column.

H. FLOW-MEASURING DEVICES

Flow measurement is extremely important in liquid chromatography, for, if the flow rate is not constant, reliable quantitative and qualitative analysis is difficult. In steric exclusion chromatography, measurement of flow and sample retention volume is important because retention volume is proportional to the logarithm of the molecular weight of the sample.

1. Bubble Timer

A bubble timer is an accurate and convenient device for measuring flow in a liquid chromatograph. An air bubble is introduced into the eluent stream and then timed while traveling between two points on a transparent section of tubing. The volume of the tube between the points must, of course, be accurately known.

A commercial bubble timer is available (Laboratory Data Control, Danbury, Conn.) which gives a digital readout of the time required for a bubble to travel between two points on a calibrated glass tube. A timing circuit is turned on and off automatically as the bubble passes between two photosensitive elements. Time is then converted to flow rate by using the known volume of the glass tube. An advantage of this type of flow-measuring device is that it does not have to be calibrated for each solvent. A precision and accuracy of about 1% make this device suitable for most applications in liquid chromatography.

2. Volumetric Methods

A volumetric procedure in which the mobile phase is collected for a known period of time in a calibrated vessel is also often used to measure flow. This method is convenient because volumetric equipment can be kept near, or even built into, the chromatograph. The siphon counter (Waters Associates, Framingham, Mass.) is an automated volumetric device that is extensively used in steric exclusion chromatography and gives a semicontinuous record of flow on the recorder chart. A small-volume vessel (usually 5 ml) fills with mobile phase until it spills over into a drain line and empties by a siphoning action. The liquid passing down the drain line triggers a photosensitive circuit which superimposes a spike on the recorder trace (see Figures 10.24 and 10.25). This type of flow-measuring device is precise and accurate to at least $\pm$1%.

3. Flowmeter

The most convenient but least accurate device for flow measurement is the flowmeter. Flowmeters use the frictional drag of the mobile phase to lift a ball vertically in a calibrated length of graduated-bore glass tubing and to give a continuous indication of flow. This method, which has found wide acceptance in gas chromatography, is not as attractive for liquid chromatography because calibration is required for each mobile phase on account of the large differences in solvent viscosity. It is convenient, however, to include a flowmeter in a chromatograph because it is the only device that gives a semiquantitative indication of flow in the system at a glance. If the flowmeter is kept free of oil and dirt, it indicates flow with a precision and accuracy of about $\pm$5%.

4. Gravimetric Methods

The most accurate method for determining flow is a gravimetric procedure in which the mobile phase is collected for a known period of time and then weighed. This method is rarely used in practice, however, because it is time-consuming and does not provide a continuous record of flow. Gravimetric methods are far more precise and accurate than is the ability of a pump to deliver liquid, so they are most frequently used in design and development for the careful evaluation of hardware.

I. GRADIENT ELUTION

Gradient elution, or solvent programming, has been used extensively in adsorption and ion-exchange chromatography and should also find considerable application in

partition chromatography with the new permanently bonded
chromatographic packings (17,18). Gradient elution in
liquid chromatography is closely analogous to temperature programming in gas chromatography and is generally
more powerful than either temperature or flow programming
for controlling capacity ratios in liquid chromatography.
With the gradient elution technique, mixtures containing
compounds with widely different chemical structures can
be totally eluted in a short period of time. A comprehensive review of gradient elution has been published by
Snyder (19).

Several questions should be answered before selecting
an apparatus for gradient elution. Should the gradient
device be versatile enough to produce gradients of any
shape, including step gradients? If a single gradient
form is all that is required, the apparatus can be
greatly simplified. Will a gradient device be compatible
with the detection system to be employed? Many detectors, especially those which measure a bulk property of
the mobile phase, are difficult to use with a gradient
device. Does the pump limit the type of gradient that
can be generated? Will an additional pump be required?
When selecting a pump for liquid chromatography, one
should consider whether the instrument will be used for
gradient elution experiments.

Gradient systems can be divided into two types: those
in which the solvents are mixed at atmospheric pressure
and pumped to the column, and those in which the solvents are pumped into a mixing chamber at high pressure
before going to the column.

1. Low-Pressure Gradient Mixers

Gradient systems in which the solvents are mixed at
atmospheric pressure have an advantage in that they are
simple to use and can be relatively inexpensive. A
single high-pressure pump is required to transport the
gradient from the vessel in which the mixing takes place,
to the column. The pump should be a reciprocating type,
such as the piston or the diaphragm model described in

Section 2.C, with a small internal volume so that mixing
cannot occur in the pump chamber and cause a step
function rather than a continuous effect.

Many methods have been reported in the literature
(19) for generating various gradients in a vessel at
atmospheric pressure. The simplest and most common of
these are the so-called exponential devices. In an
exponential device, the secondary, or strong, solvent
flows under low pressure or by gravity from a secondary
reservoir into a well-stirred primary reservoir while
the mobile phase for the chromatographic experiment is
drawn from the primary reservoir. Various gradient forms
can be generated by changing primary or secondary solvent composition, secondary solvent delivery rate, and
relative reservoir volumes. Although these devices are
usually referred to as exponential, linear gradients
(exact and approximate) and step gradients can also be
generated. In addition to the simple exponential devices, there are a number of commercially available
instruments (Instrumentation Specialties Co., Lincoln,
Neb.; LKB Instruments Co., Rockville, Md.; etc.) for generating gradients which must then be transported to the
chromatographic column via a high-pressure liquid pump.

2. High-Pressure Gradient Mixers

Gradient devices which mix the solvents together
under pressure have an advantage in that any type of
gradient can be generated simply by programming the delivery of each pump. Although some pumps are more
suitable than others, most commercially available pumps
can be adapted for flow programming. Two of the easiest
to flow-program are the screw-driven syringe-type pump
and the pneumatic pump. When two solvents are mixed at
high pressure, the design of the mixing chamber is very
important. The mixing chamber should be well swept and
should provide efficient mixing of the two solvents but
still be small enough in volume so that there is a short
time between the onset of mixing and the appearance of
the gradient at the column inlet.

Another advantage of this type of gradient system is
that it can be used effectively to evaluate mobile-phase
compositions quickly for optimizing a separation under
static conditions. Mobile-phase changes can be made
quickly by simply altering the relative delivery rate
from the two reservoirs and keeping it constant. In
this manner, many constant compositions can be evaluated
in a short time with very little solvent waste. Figure
2.14 shows the results of using a gradient accessory

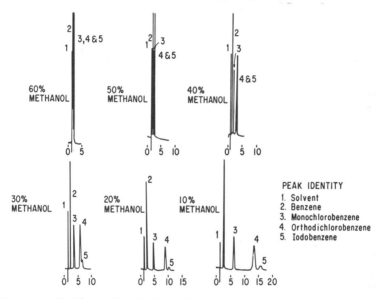

Figure 2.14. Optimization of separation by using
a gradient accessory to make rapid step changes
in mobile-phase composition. Column: 1 m x 2.1
mm i.d., precision-bore stainless steel; packing:
1% Permaphase® ODS; mobile phase: methanol-
water; sample: 5 µl of halogenated benzenes in
isopropanol; detector: UV photometer at 254 nm;
temperature: 60°C; column inlet pressure: 1200
psig; flow rate: 2 ml/min.

(E. I. du Pont de Nemours, Instrument Products Div.,
Wilmington, Del.) to make step changes in mobile-phase
composition quickly and to optimize a separation of some

halogenated benzenes. With the gradient accessory, the changeover to a new mobile-phase composition requires less than 5 min. Once a separation has been optimized in this manner, it can be set up for routine analysis with a single reservoir containing the optimum solvent mixture and a single pump for solvent delivery.

3. Commercial Gradient Accessories

Most commercial manufacturers of high-speed liquid chromatographs offer gradient elution as an accessory to their instruments. Schematic diagrams of these gradient systems are shown in Figure 2.15. Figure 2.15a shows the gradient device used on the Varian LCS-1000, and Figure 2.15b the one supplied with the Waters ALC-100. Figure 2.15c shows the gradient system used with the Nester-Faust 1200, and Figure 2.15d the gradient accessory available with the Du Pont 820. The gradient systems in Figures 2.15a and 2.15b use low-pressure mixing chambers and deliver the changing liquid to the column by means of a low-volume reciprocating pump.

The gradient devices in Figures 2.15c and 2.15d mix the solvents at high pressure. The former requires two high-pressure screw-driven pumps and uses the liquid chamber of the second pump as a mixing chamber. Figure 2.15d uses a single high-pressure pneumatic pump, a holding coil, and two proportioning valves to mix solvents from two reservoirs. A separation of some chlorinated benzene derivatives under static mobile-phase conditions and by solvent programming with the gradient system of Figure 2.15d is shown in Figure 2.16. Note that with gradient elution the later eluting peaks are quite narrow and can be eluted much faster without sacrificing the resolution of the earlier peaks.

Gradient elution is particularly useful with such detectors as the UV photometer, the fluorescence detector, and transport detectors. Operation of transport detectors can be troublesome, however, when a component in the gradient is not removed during the transport

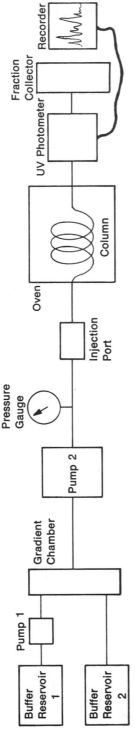

Figure 2.15. Commercial gradient elution accessories. (a) Varian LCS-1000. (Courtesy of Varian Aerograph.) (b) Waters ALC-100. [Reproduced from J. Chromatog., 43, 332 (1969) by courtesy of K. J. Bombaugh et al., and by permission of the publisher.] (c) Nester-Faust Model 1200. (Courtesy of Nester-Faust Corp.) (d) Du Pont Model 820.

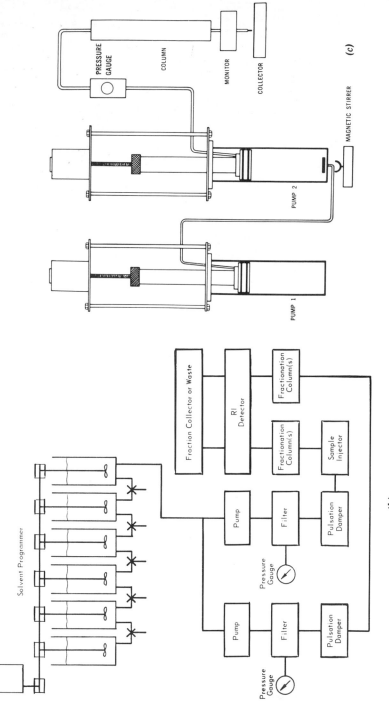

Figure 2.15 (cont)

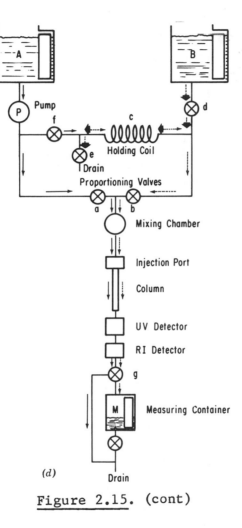

Figure 2.15. (cont)

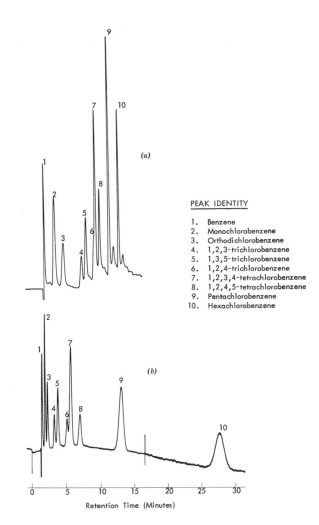

PEAK IDENTITY

1. Benzene
2. Monochlorobenzene
3. Orthodichlorobenzene
4. 1,2,3-trichlorobenzene
5. 1,3,5-trichlorobenzene
6. 1,2,4-trichlorobenzene
7. 1,2,3,4-tetrachlorobenzene
8. 1,2,4,5-tetrachlorobenzene
9. Pentachlorobenzene
10. Hexachlorobenzene

Retention Time (Minutes)

Figure 2.16. Comparison of gradient elution and constant composition for the separation of chlorinated benzenes. Chromatogram A: linear gradient 40/60 methanol-water to methanol at 8%/min. Chromatogram B: constant-composition 50/50 methanol-water. Column: 1 m x 2.1 mm i.d.; precision-bore stainless steel; packing: 1% Permaphase® ODS; sample: 5 µl of chlorinated benzenes in isopropanol; detector: UV photometer at 254 nm; temperature: 60°C; column inlet pressure: 1200 psig.

process and interferes with detection. For example, organic buffers cannot be used in the mobile phase with
the flame ionization detector. It is particularly difficult to use the so-called universal or bulk property
detectors with solvent programming.

J.   DATA ACQUISITION AND REDUCTION

1.   Recorders

A potentiometric recorder should be selected so that
it gives a faithful reproduction of the signal from the
detector. The characteristics of the detector signal
output will determine the type of recorder that is needed. Many modern detectors that have accurate bridge-
type output circuitry also have high output impedance
and require a recorder with higher input impedance. A
recorder should have an input impedance high enough so
that it draws little current in comparison with the measuring circuit.

Several other factors should also be considered when
selecting a recorder. The pen speed should be faster
than the signal from the measuring circuit; a 1-second
full-scale response is usually adequate. The ac noise
rejection should be good at 60 and 120 cps (1X and 2X
power-line frequency). The recorder should have a floating shield and terminals so that it can reproduce the
detector signal while floating several volts dc above
ground. This latter feature is important because the
output of many modern detectors is biased above ground.
If low-cost automatic integration is required, the recorder should be able to accommodate a ball and disc
type of integrator (Disc Instruments, Inc., Santa Ana,
Calif.). Full-scale deflection on the recorder should
be matched to the detector output, and there should be
enough zero offset to allow balancing of the recorder
at either end of the scale. Additional zero offset may
be desirable if small changes in a large dc signal will
be recorded.

Other features which may be important in a recorder
are number and type of pens, availability of event mark-
ers, and number and value of chart speeds. Chart speeds

of 1, 2, 5, and 10 min/in. will be used frequently. Faster chart speeds may be needed for very-high-speed liquid
chromatography, and slower chart speeds may be desirable
for steric exclusion work.

2. Digital Integrators

When high speed and accuracy are required in analytical measurements, the detector output can be coupled with
automatic data-processing equipment. Currently, the simplest and most common device found in the analytical
laboratory is the digital integrator and printer for
automatic peak area measurement. Analytical data processing is then frequently carried out manually. When large
numbers of samples are being handled or a large number
of chromatographs are being operated simultaneously, the
output from the digital integrator can be fed directly
to a computational device for rapid, automatic data processing.

3. Computers

Perhaps the most common arrangement for automatic
data processing consists of a detector coupled to a
digitizing device with time-shared access to a large
computer. Also popular are package systems, which may
include both hardware and software, and range from a
single, dedicated small computer for each instrument to
systems in which a medium-sized computer may process
data from a number of instruments in a time-shared
fashion. With the increasing availability of low-cost
dedicated computers, most chromatographers may soon have
the option of feeding their detector signal to an analog-
to-digital converter and then directly into a small computer located in the laboratory.

Hardware has also recently become available for coupling the output of a digital integrator directly to an
electronic computer-calculator. A disadvantage of this
arrangement is that the analytical calculations in chromatography require only a small part of the special
capabilities of the calculator.

Digital integrators will soon be available with computational features that will allow them to read out
directly in concentration units. In the relatively near
future, the chromatographer should be able to purchase,
for the current price of a digital integrator, a package
featuring an analog-to-digital converter and small computer which will collect and process his data rapidly and
automatically.

K.   LABORATORY SAFETY

The chromatographer should be aware that he may be
working with flammable and/or toxic liquids and should
design or select his chromatograph with safety in mind.

1. Hardware

Very little potential energy is stored up in a liquid
under pressure at ambient temperature; therefore, no
special precautions need be taken to protect the chromatographer from a sudden release of energy when a fitting
becomes loose or leaks. It should be pointed out, however, that when the system is heated so that the liquid
reaches its supercritical point, the danger of violent
explosion is present.

Nitrogen purge should be included in the system to
prevent combustion in any part of the liquid chromatograph where a flammable mobile phase can come into contact with a hot surface. An example of such a potential
danger point is a heater element or an electrical connection that can generate a spark.

Relief valves or rupture discs leading to a waste container should be included with mechanical pumps to prevent damage or spills when a blockage occurs in the
system. Constant-pressure pneumatic pumps cannot exceed
the desired pressure unless the gas regulators fail and,
for this reason, generally do not require a pressure-
relief device. Pressure gauges are also available which
activate a solenoid switch at a selected pressure. The
solenoid switch can then cut the power to a mechanical
pump or shut off the gas pressure to a pneumatic pump.

2. <u>Ventilation</u>

A liquid chromatograph should be placed in a well-ventilated laboratory. Solvent spills and leaks are inevitable, and prolonged exposure to even small amounts of certain solvents is undesirable. An "elephant trunk" positioned above the liquid chromatograph, similar to the arrangement often used with atomic absorption spectrometers, is not adequate because most solvents are heavier than air and vapors from solvent spills spread quickly along the floor. Ideally, exhaust intake should be located near the floor and close to the liquid chromatograph. Exhaust should also be 100% make-up air and should not be tied into the ventilation system of another laboratory.

3. <u>Solvent Handling and Disposal</u>

Solvents should be stored in well-ventilated cabinets, and large quantities should be located in a remote part of the building or in a separate shed. The use of safety containers to transport solvents is strongly recommended. No more than one gallon of any toxic, flammable, or otherwise dangerous solvent should be stored in the laboratory at one time.

Disposable gloves should be worn when handling toxic solvents, and spills should be wiped up immediately with strongly absorbent towels. A special trash can with a lid should be kept for any contaminated items that could be harmful if accidentally touched. The chromatographer should take the responsibility of seeing that the contents of this trash can are properly disposed of (incinerated, if possible). Nonaqueous solvents that have been used as the mobile phase should be placed in special containers that are regularly taken to an incinerator. Under no circumstances should the waste lines of a liquid chromatograph be allowed to drain into laboratory sewer systems.

4. <u>Sample Handling and Disposal</u>

Extreme care should be taken when handling toxic or otherwise dangerous samples. Information on toxicity

and other dangerous properties of chemicals can be found
in a handbook by Sax (20). Ventilated cabinets and re-
frigeration facilities will be required if large numbers
of different samples are investigated in the laboratory.
Special vials which allow syringe sampling through a
Teflon® cap (Precision Sampling, Houston, Tex.) are excellent for toxic or valuable samples which should not
be spilled. Syringe needles should be kept sheathed at
all times to prevent accidental poisoning by injection.
When toxic samples are being injected, disposable gloves
should be worn to prevent accidental skin contact if the
syringe breaks or leaks back around the plunger barrel.
Stop-flow injection may be desirable with highly toxic
samples.

Old samples and sample vials should be accumulated in
a special container and, at intervals, incinerated or
buried.

References

1. T. E. Young and R. J. Maggs, Anal. Chim. Acta., 38
   105 (1967).
2. R. E. Jentoft and T. H. Gouw, Anal. Chem., 38, 949
   (1966).
3. H. Felton, J. Chromatog. Sci., 7, 13 (1969).
4. S. T. Sie and G. W. A. Rijnders, Separation Sci.,
   2, 729, 775 (1967).
5. R. E. Jentoft and T. H. Gouw, J. Chromatog. Sci.,
   8, 138 (1970).
6. R. A. Henry, unpublished data.
7. J. J. Kirkland, unpublished data.
8. C. D. Scott, W. F. Johnson, and V. E. Walker, Anal.
   Biochem., 32, 182 (1969).
9. J. J. Kirkland, J. Chromatog. Sci., 7, 361 (1969).
10. R. E. Majors, Symposium on Advances in Chromatography,
    1970, Miami Beach, Fla., June 2-5, 1970.
11. R. P. W. Scott, D.W. Blackburn, and T. Wilkins, J.
    Gas Chromatog., 5, 183 (1967).
12. J. Kwok, L. R. Snyder, and J. C. Sternberg, Anal.
    Chem., 40, 118 (1968).

13. J. J. Kirkland, paper submitted to J. Chromatog. Sci.
14. F. W. Peaker and C. R. Tweedale, Nature, 216, 75 (1967).
15. C. D. Scott, J. Chromatog., 42, 263 (1969).
16. R. P. W. Scott, Symposium on Modern Practice of Liquid Chromatography, Wilmington, Del., April 6-8, 1970.
17. I. Halàsz and I. Sebastian, Angew. Chem. Internat. Ed., 8, 453 (1969).
18. J. J. Kirkland and J. J. DeStefano, J. Chromatog. Sci., 8, 309 (1970).
19. L. R. Snyder, in "Chromatographic Reviews," Vol. 7, M. Lederer, ed., Elsevier, New York, 1965.
20. N. I. Sax, "Dangerous Properties of Industrial Chemicals," Reinhold, New York, 1963.

CHAPTER 3

Detectors in Liquid Chromatography

S. H. Byrne, Jr.

A. INTRODUCTION

The detector is used in liquid chromatography to monitor the concentration of the solute in the mobile phase continuously as it leaves the column. The selection of a detector is an important consideration in analytical liquid chromatography.

In gas chromatography, the gaseous mobile phase has properties which are usually significantly different from those of the solutes which are passed through the system. Detectors using flame ionization, thermal conductivity, or density can be readily employed for the detection of small concentrations of solute in the carrier gas. Unfortunately, the mobile phases used in liquid chromatography have physical properties very similar to those of the solutes. Therefore, to monitor solutes in liquid chromatography one must either eliminate the solvent before the sensor (flame ionization), find solute properties for which the mobile phase does not interfere (UV absorption), or measure one of several bulk physical properties by means of careful reference compensation and temperature control (refractive index). Although there is no one best or universal detector for liquid chromatography, there is generally a best detector for a particular application. If samples range widely with respect to their physical properties, there is no choice but to have several detectors.

Two types of detection devices, bulk property detectors and solute property detectors, are available. As the solute emerges from the column, the bulk property detector measures a change in an overall physical property of the mobile phase. Refractive index, conductivity, and dielectric constant detectors are of this type. The solute property detector is sensitive to physical properties of the solute which are not exhibited to any extent by the mobile phase. Examples of this type are the UV

95

absorption, polarographic, and radioactivity detectors. To detect trace quantities, bulk property detectors must be compensated and the temperature carefully controlled. Bulk property detectors generally are less sensitive and are more prone to noise and drift because of the difficulty of controlling temperature to the required value.

The two detectors which have the widest range of applications and are the most often used are the UV absorption and the refractive index. The UV absorption system (generally at 254 nm and 280 nm) is one of the most sensitive detectors in liquid chromatography and can detect solutes in the nanogram range for samples which have moderate absorptivities. Ultraviolet absorption is easy to use and is relatively insensitive to flow and temperature. Obviously, a compound must have some absorbance at the detection wavelength if this detector is to be used.

The differential refractometer is useful for samples which do not absorb in the ultraviolet. It is fairly simple to use and can detect solutes in the microgram range. The differential refractometer is, however, sensitive to changes in temperature and flow.

When evaluating potential detectors for a particular application, it is often necessary to delve into a forest of specifications in order to determine the applicability of a detector. Each specification is generally quoted differently, and all must be reduced to common terms. Unfortunately, for most of the samples that are of interest, no reference values for many of the properties used in detectors are available. To gain the needed information on performance, one must either find a similar application in the literature, rely on the applications laboratories of the various manufacturers, or actually test the sample with the detector under consideration.

The primary detector parameters are the following:
1. Noise
    (a) Short-term noise
    (b) Long-term noise
    (c) Drift

2. Sensitivity
   (a) Absolute sensitivity
   (b) Relative sample sensitivity
3. Linearity

Other operational parameters which should also be known or investigated include:

- Base line drift caused by changes in flow
- Sensitivity changes with respect to flow
- Sensitivity of the detector to changes in room temperature
- Sensitivity of the detector to changes in mobile-phase temperature
- Dead volume and band spreading of the cell and inlet tubing
- Ease of operation

Table 3.1 shows relative parameter values for several detectors.

1. Noise

Noise is defined as the variation of the output signal of a detector which cannot be attributed to the solute itself passing through the cell. Noise can arise from the electronics in an instrument, temperature fluctuations, line voltage changes, flow changes in the cell, and other causes. When comparing instruments, it is important to note the experimental conditions under which the detector specifications apply.

Detector noise takes three forms. Short-term noise is the variation which, on the recorder trace, tends to widen the width of the trace and which appears as "fuzz" on the base line. Long-term noise is the variation of the recorder tracing which appears as "peaks and valleys" on the base line. Long- and short-term noise can be periodic or random. Short-term noise increases the minimum detectable limit, and long-term noise makes the finding and identifying of small peaks difficult. Drift is the steady movement of the base line up or down scale. Drift tends to camouflage noise, and no noise determination can be made unless the drift is small in relation

Table 3.1

Detectors Used in Liquid Chromatography ([13-16])

Type of Device	Units	Upper Limit of Linear Dynamic Range	Full-Scale Sensitivity at $\pm$1% Noise	Sensitivity To Favorable Sample	Sensitivity Change with Flow	Temperature Sensitivity	Cell Volume, μl
UV absorption	AU	2.56	0.005	5×10^{-10} g/ml	Ind. of flow	negligible	10
Refractive index	RIU	10^{-3}	10^{-5}	5×10^{-7} g/ml	Ind. of flow	10^{-4}°C	3
Polarographic	μA	2×10^{-5}	2×10^{-8}	10^{-9} g/ml	Prop. to flow	1.5%/°C	10
Adsorption	°C	10^{-1}	5×10^{-3}	10^{-9} g/sec	Prop. to flow	5×10^{-5}°C	9
Fluorimeter		(one order of magnitude)		10^{-9} g/ml	Ind. of flow	--	10
Flame ionization	A	10^{-8}	10^{-11}	10^{-8} g/sec	Prop. to flow	--	--
Conductivity	μmho	1000	0.05	10^{-8} g/ml	Ind. of flow	2%/°C	1.5

to the magnitude of the noise. Noise is generally quoted
as a percentage of full-scale value at a specified sensitivity or as a specific peak-to-peak voltage.

2. Sensitivity

The absolute detector sensitivity is the total change
in a physical parameter for a full-scale deflection of
the recorder at maximum detector sensitivity with a defined amount of noise. To compare detectors, sensitivities must be quoted at equal noise magnitude. Sensitivity can also be quoted at the value equal to the noise
or equal to twice the noise. It is usually assumed that
the height of a peak has to be at least twice the noise
in order to be detected. Thus, if the detector noise
band is 2%, the minimum detectable sample would cause
the base line to move 4%. To determine the absolute
sensitivity, a standard solution with known properties
is usually injected directly into the detector.

The relative sample sensitivity of a detector is the
minimum concentration of solute which can be detected.
As stated above, this is usually considered to be the
concentration which causes a peak that is double the value of the noise.

To determine the relative sensitivity for a particular sample, the detector should be installed in a liquid
chromatographic system with pump, injection port, column,
etc. The system is held at constant flow rate, temperature, and other operating conditions. The minimum detectable limit can then be obtained with a sample of known
concentration by injecting solutions of decreasing concentration into the chromatograph. The responses of
many bulk property detectors depend on the solvent as a
reference value (i.e., refractive index), and, therefore,
the sensitivity will change with a change in solvent.
The column and dead volume in the chromatographic system
will cause peak band broadening, which increases the minimum solute size that can be detected. The sensitivity
of some detectors depends on flow (e.g., polarographic),
and this should be recognized.

The average concentration in the cell at the minimum
detectable limit can be found approximately by dividing
the original solute weight by the volume of mobile phase
passed during the peak (peak width time multiplied by the
flow rate). The concentration at the peak is approximately double the average concentration. To ensure that
broadening of the peak is minimized as it passes through
the detector, the cell volume should be considerably less
(about two orders of magnitude) than the volume of mobile
phase which contains the peak.

3. Linearity

For a detector to function as a quantitative analytical instrument which measures solute concentration of the
effluent of the column, it is highly desirable that the
signal output be linearly proportional to the solute con-
concentration. All detectors used in liquid chromatography approach this goal, but none is perfectly linear
over its entire range of detectability. (Table 3.1 indicates the upper limit of linearity for each detector.)
There are two commonly used measures of linearity.

   1. Using the equation (11)

$$D = A_R c^x \tag{3.1}$$

where D = detector output, A_R = proportionality constant
called the response factor, c = concentration of solute
in the mobile phase, and x = exponent, one finds that a
truly linear detector will have the constant $x = 1$ for
all values of c (i.e., $D = A_R c$). For all detectors in
use, however, x approximates unity over only a small
range. In practice, linearity can be assumed only for
values of c for which the value of x falls in the range
from 0.98 to 1.02.

   2. A measure of linearity is the maximum percentage
deviation from linearity over a range expressed as a percentage of full-scale deflection.

Linearity can be determined by injecting progressively
larger quantities of solute onto the column and measuring
the peak heights or peak area of the eluted peaks. All
operating conditions should remain constant, and tripli-
cate runs should be made. At the end of the sequence

of injections, the results should be plotted. To determine x in equation 3.1, it is necessary to plot the log of the peak area (or height) against the log of the concentration. For a perfectly linear instrument, the graph will have a slope of 1. If the instrument is not perfectly linear, the slope of the line at any point is equal to the value of x. Computer programs are readily available for the calculation of x and should be used if available, since results will be much more accurate (1). If the detector is nonlinear, appropriate correction factors can be applied to determine concentration.

There are many general considerations which should be noted when evaluating and using detectors. All detectors must be connected to the outlet of the chromatograph column; therefore, dead volume should be kept to a minimum by using minimum-bore capillary tubing and short coupling distance. When capillary tubing is used, a frit is required on the column output to prevent column packing particles from leaving the column and clogging the tubing. Flow into the cell cavity should be smooth, with no unswept areas. Cell windows or interfaces should be kept clean and should also be well swept with mobile phase. Cell volume should be kept to the minimum compatible with the requirements of the particular detector.

Bubbles cause an intolerable amount of noise in most liquid chromatographic detectors. All fittings should be airtight, since air can "diffuse" into the mobile phase with no external signs of liquid leakage. The mobile phase can be degassed either by vacuum or by boiling. Back-pressure applied on the cell outlet within the limit of the rated pressure of the cell will tend to keep bubbles in solution and will help to dissolve bubbles which cling to the cell. Dirt or an immiscible phase in the cell can produce similar noise on the recorder.

The recorder should have an input impedance which exceeds the output impedance of the detector by a factor of at least 100 and should have the correct voltage span. The detector should have zero suppression or offset which will compensate for background from the mobile phase.

Short-term noise is generally suppressed in the output circuit of the detector electronics. If the suppression is too great, sharp peaks will be delayed and flattened. In high-speed chromatography, peaks will not be altered if the overall system time constant does not exceed about 0.5 sec. (To determine the time constant of an instrument, a step change in concentration should be introduced into the cell. The time constant is the time needed for the recorder to reach 63.2% of its final value.)

If detectors are placed in series, the second detector should be the one with the greatest dead volume. A common combination is the UV absorption-refractive index pair. Parallel detectors can be used in some instances; however, sensitivity is lost because of flow splitting. Also, quantitative measurement is difficult because the relative proportion flowing to each detector will vary with changes in mobile-phase viscosity. Solute peaks can further alter these proportions as they pass the splitter.

B. ULTRAVIOLET ABSORPTION DETECTOR

The detector most widely used in high-speed liquid chromatography today is the UV absorption device. It is relatively insensitive to temperature and flow changes and has a high sensitivity to many samples. Detectors of this type with sensitivities of 0.005 AUFS (absorbance units full scale at $\pm$ 1% noise) are available. With these high sensitivities, samples with relatively low absorptivities can be detected. It is possible to detect samples as small as 3 ng under variable conditions.

Monochromatic instruments are available from Du Pont, Nester-Faust, Waters, Laboratory Data Control (LDC), Chromatronics, Beckman, and Varian. (A complete list of manufacturers' names and addresses appears at the end of this chapter.) All of these units are single- or dual-wavelength photometric devices which operate on the same basic principle with minor variations. Spectrophotometric devices with a wide range of wavelengths are not as useful for high-speed chromatography because of their relatively lower light energy and high cost. A spectrophotometer is available from Gilford Instrument which will detect samples as small as 1 μg.

The basic layout for a UV photometer is pictured in Figure 3.1. Light from the source is collimated by lens A, split into two beams by the semitransparent mirror B, and focused on the cells by lenses C and D. Light from

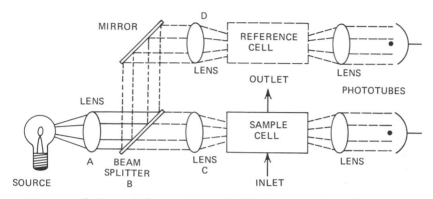

Figure 3.1. Schematic of UV absorption detector.

the cell is again collimated and focused on the photocells. If the photocell is sensitive to energy at wavelengths other than the one of interest, it is necessary to put a blocking filter in the measuring channel.

Typical sources are low-pressure mercury (254 nm), medium-pressure mercury (280 nm), and phosphors (280 nm reradiated from a phosphor which is exposed to a 254-nm source).

To make the photometer output linear with concentration, it is necessary for the output of the photocells to be fed to log amplifiers (Figure 3.2). The output from the log amplifiers is subtracted, amplified, and fed to a recorder. Some instruments do not use the log amplifier but rely on the response of the photodetector to approximate the log function. This approximation is good only to about 0.009 absorbance (2).

Cells of two different types have been used in UV photometric detection. The classical type is of Z configuration shown as in Figure 3.3. The mobile phase comes in one end, sweeps the window, and goes down the cell cavity and out the other end. To minimize noise

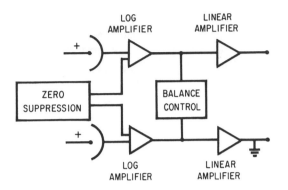

Figure 3.2. Electronics of UV absorption detector.

and drift due to flow variations or fluctuations, an H
cell (Figure 3.4) has also been used (3). The mobile
phase flows into the center of the cell, splits, and goes
in opposite directions down the bore of the cell. At
both ends, the mobile phase sweeps the windows and is re-
combined in the upper bore.

Although a reference cell can be used if the absorbance of the mobile phase is changing, it is very difficult to match the two streams exactly in flow and hold-up
volume. Frequently, the reference flow cell causes more
drift and noise than it corrects; however, it also can
be used to balance out excessive background absorbance
in the mobile phase.

Sample solvents and other nonabsorbing compounds sometimes produce a signal in the UV detector because of the
fact that there are multiple optical paths from the re-
flections off the sides of the narrow-bore cell. These
reflection intensities are altered when the refractive
index of the mobile phase in the cell changes. In some
designs, the light entering the cell is collimated to
reduce this effect. Since the flow inside the cell is
not uniform, turbulence is generated which tends to
scatter the light in the cell. As the viscosity or
velocity of the stream changes, the flow pattern will
change and may affect the signal.

The UV photometer is an ideal detector for use with
gradient elution. Many common solvents which are useful

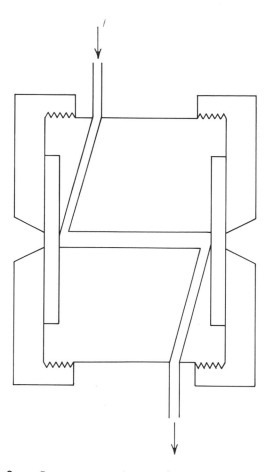

Figure 3.3. Cross section of Z flow cell (14).

in liquid chromatography have low absorptivities at the
detector wavelengths, and these can be programmed without
causing excessive detector base line drift.   In ion-
exchange chromatography, pH and ionic strength can be
varied without changing the detector base line, provided
the ion selected does not absorb UV (as illustrated in
Chapter 8).

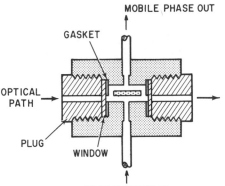

<u>Figure 3.4.</u>   Cross section of H flow cell (<u>3</u>).

C. <u>FLUORIMETER</u>

The fluorimeter detector measures the fluorescent
energy from a solute which has been excited by UV radiation. Many compounds absorb UV light and fluoresce at
higher wavelengths; often the fluorescent wavelength is
in the visible region. Many biological compounds, including certain metabolites, many drugs, amino acids,
amines, vitamins, and steroids, can be detected with a
fluorimeter. Fluorescent derivatives of many compounds
also can be made.

In the operation of a fluorimeter, the light from a
UV source is filtered and focused on the cell, which can
be made in either the right-angle (Figure 3.5) or the
straight configuration (Figure 3.6). The initial wavelength is then blocked by a filter, and the intensity of
the emitted energy is measured by a photocell. A reference cell can be used to cancel out background fluorescence radiated from the mobile phase. When the fluorimeter is used, the mobile phase should not absorb radiation
at the initial or the fluorescing wavelength. With
strongly fluorescing compounds such as quinine sulfate,
the sensitivity can be as high as 10^{-9} g/ml. Usually
the UV absorption detector is as sensitive as the fluorimeter and is more useful for quantitation. However, the
fluorimeter is very useful as a specific detector to

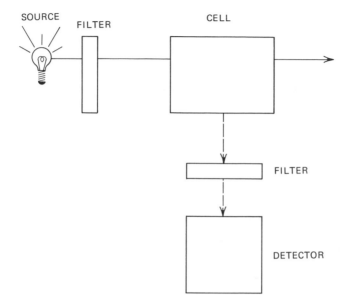

Figure 3.5. Right-angle fluorimeter detector.

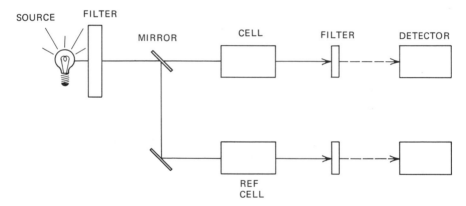

Figure 3.6. Straight fluorimeter detector.

identify particular compounds. A dual-beam fluorimeter for high-speed chromatography is commercially available from LDC.

D. RADIOACTIVITY DETECTOR

Although not widely employed, the radioactivity detector is a very useful specific detector when labeled species are available or can be generated. The technique
has a wide linear range and is insensitive to solvent
changes. Weak beta emitters such as ^{35}S, ^{14}C, and ^{3}H
have all been detected. No commercial detectors for the
continuous monitoring of eluent streams in high-speed
liquid chromatography are currently available; however,
devices may be purchased for use as detectors in less
demanding applications. The flow cells of commercially
available units are large (greater than 1 ml). However,
when the resultant band spreading can be tolerated, this
type of detector is quite useful in many specific problems. Packard Instruments manufactures a unit which can
be adapted to high-performance liquid chromatography.

In this method of detection, scintillators (such as
calcium fluoride, anthracene, t-stilbene, and diphenyl-
stilbene) are placed in direct contact with the effluent
of the column. As the radiation is emitted, light pulses
are triggered in the scintillant and detected by a
photomultiplier. The output from the photomultiplier is
fed into a rate meter and recorded.

The efficiency of a scintillation system is defined as
the percentage of the total number of disintegrations
which result in a measurable pulse of light. Three types
of scintillation cells have been used:

1. Tube cells - plastic scintillation tubing is formed
into a spiral (lengths up to 100 cm; internal diameters
of about 1.7 mm). Efficiencies of 5.7% for ^{14}C have been
reported (4).

2. Packed cells - cells are packed with scintillator
crystals or particles exposed directly to the column
effluent. Solutes of amino acids, peptides, proteins,
and glucose have been reported with efficiencies of 55%
for ^{14}C and 2% for ^{3}H (4).

3. Mixing systems - upstream of the cell a stream of
liquid scintillator is mixed into the column effluent.
The scintillator must be miscible with the mobile phase.

Efficiencies of 70% for ^{14}C and of 30% for ^{3}H have been reported (5).

Optical density changes of the effluent, contamination or quenching of cell packing or tubing by the mobile phase or solutes, chemiluminescence, flow rate changes, cross-contamination of double-labeled samples, and background radiation - all these factors can affect the performance of this detector.

Since diffusion occurs slowly in liquids, radioactivity detectors offer the unique possibility of stopping flow in the cell and counting a particular peak for a period of time to improve sensitivity.

E. POLAROGRAPHY

Polarography is an electrochemical method which measures the current between a polarizable and a nonpolarizable electrode as a function of applied voltage. When this approach is used as the basis of a detector in liquid chromatography, a constant voltage is normally applied between the electrodes, which results in either oxidation or reduction of the solute. The current which flows is then measured continuously and recorded as a function of time. Metal ions, nitro compounds, DDT isomers, amino acids, alkaloids, aldehydes, ketones, and inorganic anions have all been detected in column effluents by this technique (6).

Three detectors with different polarizable electrodes have been reported in the literature.

1. The dropping mercury electrode (6) (Figure 3.7) uses a polarizable electrode which is constantly renewed by an incoming stream of mercury. The flow from the column impinges on the mercury electrode as it enters the cell. The nonpolarizable electrode is formed by the pool of mercury at the bottom of the cell. The dropping mercury electrode is extremely sensitive to dissolved oxygen and must have high electrical damping to minimize the noise caused by the change in current as the mercury forms and drops from the electrode.

2. Another polarographic detector, which uses a carbon-impregnated silicone rubber electrode (7), is shown

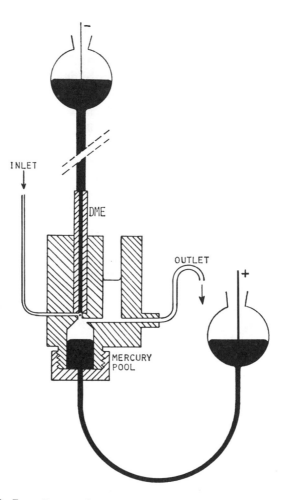

Figure 3.7. Dropping mercury polarographic detector (6).

in Figure 3.8. The column effluent contacts the polarizable electrode in the small space below the electrode formed by the silicone O-ring. Mercury is used to assure good electrical conduction between the wire electrode and the silicone rubber. The nonpolarizable electrode is made of platinum and is located downstream.

3. Platinum electrodes are sometimes used, but are contaminated easily and must be regenerated frequently.

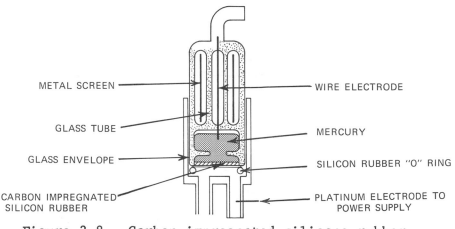

Figure 3.8. Carbon-impregnated silicone rubber polarographic detector (7).

No commercial polarographic detectors are presently available for liquid chromatography. This type of detector is difficult to use quantitatively, since peak areas vary with mobile phase, column, flow rate, and voltage. However, since each solute has a different optimum mobile phase and voltage, the polarographic detector can be quite selective. The mobile phase must have a high conductivity and is usually a mixture of water with salts or organic solvents. Care should be taken in the construction of the cell to reduce the dead volume and to keep the other electrode as close as possible to the polarizable electrode.

F. HEAT OF ADSORPTION

The heat of adsorption detector (8,9) has a wide range of applicability in liquid chromatography and is quite general in application. Unfortunately, however, several serious inherent disadvantages limit the usefulness of this device. Its basis for operation is the change in temperature associated with the adsorption phenomenon. As a solute is adsorbed onto a surface, a temperature increase takes place. Conversely, a decrease in temperature occurs as the solute is desorbed. These small changes in temperature form the basis of the heat of adsorption detector.

Temperature changes can be measured within the column, at the column exit, or in a separate cell packed with column packing. Measurement is made either by thermocouples or by thermistors. The external cell (Figure 3.9) is most often used, since it offers the greatest

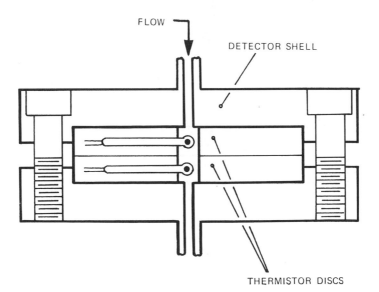

Figure 3.9. Heat of adsorption cell (8).

latitude in cell configuration and provides minimal band spreading. A second temperature sensor (reference) is needed to compensate for temperature changes in the solvent. This reference sensor usually is placed upstream from the sensing cell in a glass bead packing to avoid temperature effects which are transmitted in the carrier from the packing in the measuring cell. Commercial instruments of this type are available from Varian and Nester-Faust.

Since thermistors with suitable amplification are capable of measuring $10^{-5}°C$, temperature control of the

cell and incoming mobile phase is very important. Positive temperature control to this degree is impractical;
therefore, a large water bath is generally employed to
keep the rate of change of temperature (and therefore
drift and noise) at an acceptable value. Both the cell
and a length of inlet tubing used to bring the mobile
phase to constant temperature are immersed in the bath
so that they will attain the same temperature. Temperature control of the bath should be $\pm 0.003°C$.

The resulting peak due to the adsorption-desorption
is an unsymmetrical S-shaped curve, as shown in Figure
3.10. Therefore, the principal use of the heat of

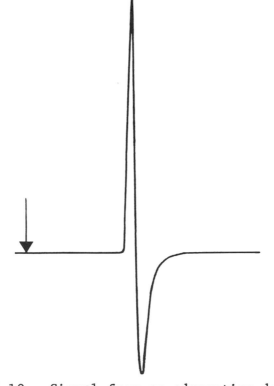

Figure 3.10. Signal from an adsorption detector.

adsorption detector is for the qualitative detection of
solutes in scouting work. Unresolved and closely spaced

peaks are not easily resolved because of the double peak.
With this detector, peak height is a function of concentration but calibration is not constant because of contamination or deactivation which may occur in the cell
packing.

The heat of adsorption detector is also flow sensitive
as a result of the self-heating of the thermistors. In
organic solvents this effect is about 2 x 10^{-4}°C/cc/min.
Thus, gradual changes in flow can cause drift, and pulsations from a constant-volume reciprocating pump produce
unacceptable noise unless the pressure fluctuations are
well damped. Gradient elution can be used only at low
sensitivity because of base line drift caused by the temperature change associated with the changing of mobile
phases.

G.   SOLUTE TRANSPORT DETECTORS

This type of detector operates on the principle of
removing the volatile carrier before the detection process. A sample of the column eluent is continuously fed
onto a suitable transport system, such as a moving belt,
wire, coil, or chain. The volatile mobile phase is evaporated, and the residual nonvolatile solute adhering to
the transport system is detected directly. The advantage
of this type of system lies in the fact that its detection capability is entirely independent of the methods
of chromatographic development. Since the mobile phase
is continually removed from the transport system before
any attempts are made to detect the solute, the chemical
nature, temperature, and other characteristics of the
mobile phase have no effect on the detection system.
This method of detection is limited to relatively nonvolatile solutes since more volatile compounds will be
evaporated with the mobile phase.

Commercial solute transport detectors based on the
flame ionization detector are manufactured by Phillips,
Nester-Faust, Pye Unicam, Packard, and Nuclear-Chicago.
In this type of detector, wire (or other transport system) is fed continuously through an oven at about 750°C,
where it is cleaned by pyrolysis. Subsequently, the

clean wire passes through the column eluent and is coated
with a thin film of the mobile phase containing any solute present. After coating, the wire passes through an
evaporator chamber to remove the solvent and then into
another oven, where the residual solute is pyrolyzed. A
diagram of the system manufactured by Phillips is shown
in Figure 3.11. It is seen that the gas flows are

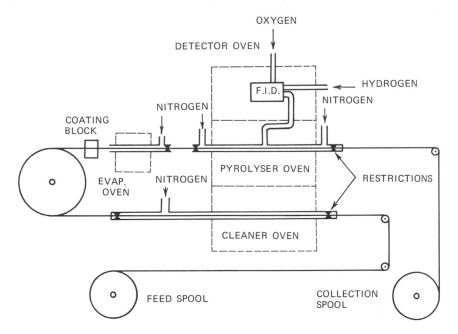

Figure 3.11. Schematic of wire transport detector. (Courtesy of Phillips Electronics Instruments.)

arranged so that the pyrolysis products from the wire
pass into a hydrogen flame detector. The ion current
produced from the combustion of the pyrolysis products
in the flame detector provides the detection signal.
This detector has a linear response with a response
index lying between 0.96 and 1.04 for a concentration
range of about two orders of magnitude but has a varying
sensitivity, depending on the nature of the pyrolysis
products. Whereas this detector is very sensitive to

high-molecular-weight hydrocarbons, it is relatively insensitive to carbohydrates.

Recently an alternative method of detection based on
the same principle has been described. This detector
arranges for the solute to be converted to carbon dioxide,
which is then mixed with hydrogen, converted to methane,
and detected by a flame ionization detector (10); a diagram is shown in Figure 3.12. The solute deposited on
the wire is burnt at 850°C in a quartz tube through which

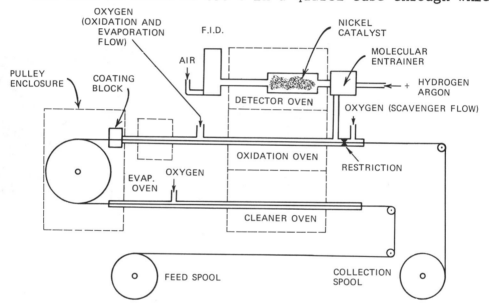

Figure 3.12. Schematic of modified wire transport detector (10).

air or oxygen is flowing. The carbon dioxide produced
is aspirated into a hydrogen stream by means of a molecular entrainer (a venturi device), which then passes over
a catalyst at 350°C. The catalyst converts the carbon
dioxide to methane, which then passes to a flame ionization detector. This system has a predictable response
related to the carbon content of the solute being detected. Furthermore, it has a response index of 0.98-1.02
over a concentration range of five orders of magnitude.

The transport detector has a linear response over a
wide range of solute concentration, and its absolute sensitivity is about 2-3 µg/ml (11). The chief disadvantages
of the solute transport detector in its present form are
its large size and relative insensitivity. The relatively poor sensitivity is due mainly to the fact that the
wire picks up only a small part of the column effluent.
Some of the noise in the detector can be attributed to
uneven coating of the wire. Seal leaks at the ends of
the ovens can also cause loss of solute. Wire speed and
gas flow rate must be carefully controlled to obtain good
results.

H. DIFFERENTIAL REFRACTOMETER

The differential refractometer detector continuously
monitors the difference in refractive index between a
reference mobile phase and an analytical mobile phase as
elutes from the column. Samples as small as 3 µg/ml can
be detected under favorable conditions.

The refractometer is a versatile instrument for measuring a wide range of samples but has several diasdvantages.
The most important is that refractive index is extremely
sensitive to temperature; temperatures must be held to
$\pm$ 0.001°C if sensitivities of 10^{-5} RIUFS (refractive
index units full scale at $\pm$1% noise) are to be used.
Refractomers are sensitive to flow changes, particularly
at elevated temperatures. The sensitivity to a particular solute is, of course, dependent on the difference
between its refractive index and that of the mobile phase.
It is not possible to use this method of measurement in
systems which use gradient elution except at low sensitivity or in the rare instance where the refractive
indices of the solvents can be matched.

Several commercial instruments are available. They
operate on two different principles: the deflection
type (Waters) and the Fresnel type (LDC, Du Pont, Varian,
Nester-Faust). Sensitivities of 10^{-5} RIUFS are common.

1. Fresnel Refractometer

The Fresnel refractometer is based on Fresnel's law

of reflection, which states that the percentage of light
reflected (or transmitted) at a glass-liquid interface
is proportional to the angle of incidence and the refractive indices of the two substances. To obtain maximum
sensitivity from this type of instrument, it is necessary
to operate at an angle slightly less than the critical
angle at the liquid-glass interface. Noise and temperature effects are minimized through the use of a differential measurement of refractive index. Optical misalign-
ments and chance variations in the intensity of the light
striking each stream are avoided by illuminating both
cells with a single lamp whose rays are collimated and
focused through a common set of components.

A schematic of the optical system is shown in Figure
3.13. Light from source lamp SL passes through source

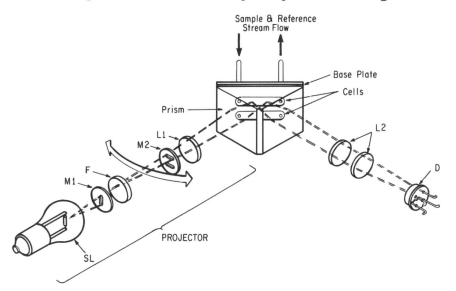

Figure 3.13. Fresnel refractometer. (Courtesy
of Du Pont Instruments.)

mask M1, infrared blocking filter F, and aperture mask
M2 and is collimated by lens L1. Mask M2 defines two
collimated beams which enter the cell prism and impinge
upon the glass-liquid interfaces as shown. Sample and
reference liquid chambers are formed by a Teflon® gasket

which is clamped between the cell prism and the stainless steel base plate. Cell volume is about 3 μl. Components SL through L1 are mounted on a separate optical
bench which can rotate about a pivot point to change the
angle of incidence. Light, which is transmitted through
the two interfaces, passes through the thin liquid film
and impinges on the surface of the cell base plate. Lens
L2 focuses this light on the dual-element photodetector,
D. The electrical signal from the detector is then amplified and relayed to the recorder. Since the Fresnel
effect occurs through the liquid-glass interfaces, the
glass must be kept clean in order for the detector to be
used.

2. Deflection Refractometer

When a beam of light is passed through a deflection
refractometer cell which has two liquids of different
refractive indexes, the beam is displaced by an amount
proportional to the refractive index difference between
the two liquids. In Figure 3.14, the light emanating

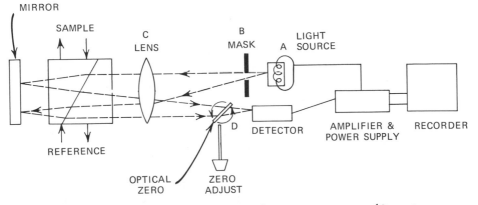

Figure 3.14. Deflection refractometer. (Courtesy
of Waters Associates.)

from source A and limited by mask B is collimated by lens
C and passed through the cell. The cell has a reference
and a sample compartment, the two being separated by a
diagonal piece of glass. As the composition of the mobile phase changes, the light is deflected. (The

reference cell compensates for changes in the refractive
index of the mobile phase caused by temperature fluctuations.) The light is then reflected from the mirror back
through the cell and again deflected. The lens focuses
the light on the position-sensitive photodetector, which
produces an electrical signal proportional to the position of the light image. This signal is amplified and
relayed to the recorder. The optical flat (D) is used
to displace the light and to balance the instrument.

   Although this system has the advantage of a wide range
of linearity, it is quite sensitive to the relative motion of the light cell and the detector. The cell is not
as small or as cleanly swept as the Fresnel type.

3. Heat Exchangers in Differential
Refractometer Detectors

   Since the refractive index of a liquid is inherently
sensitive to temperature (approximately 10^{-4} RIU/°C), it
is of primary importance that the analytical and reference streams be at the same temperature. Noise and drift
will occur in both types of refractometer if the temperature of either of the two streams changes. The temperatures of the analytical and reference streams should be
equalized before entering the refractometer by passing
the inlet tubing of both streams through a large thermal
mass of highly conductive metal or a well-controlled water
bath. The two streams should be brought to the same constant temperature by exposing each to the same thermal
environment. The design of the heat transfer system
should be such that there is minimum dead volume in the
analytical stream to minimize band spreading (12). The
tubing should be as small as possible (0.010 in.) and
should be long enough to give the liquids time to equili-
brate. The flow should be held constant, since flow
variations will also cause temperature fluctuations.

I. MISCELLANEOUS DETECTORS

   Several other detectors have been used which rely on
a physical property of the mobile phase-solute combination. All of these "bulk" property detectors are subject

to temperature and flow problems and require rigorous
temperature control and other compensation. Generally,
they are not sensitive enough to detect trace quantities
in high-speed chromatography. Most do not work well with
gradient elution.

1. Conductivity

Conductivity can be measured with either dc or high-
frequency ac detection systems. The electrodes of the
dc system polarize during operation, causing nonlinearity
and drift. Although ac conductance measurements using a
simple bridge detection network reduce the polarization
effect, they include the reactive (out-of-phase) component caused by the dielectric constant of the solvent.
This effect can be overcome by using a phase-sensitive
detector which measures the in-phase or conductive current from the cell.

Electrodes from the cell are made of platinum, stainless steel, or other inert metal. A diagram of a sample
cell is shown in Figure 3.15. The temperature of such a

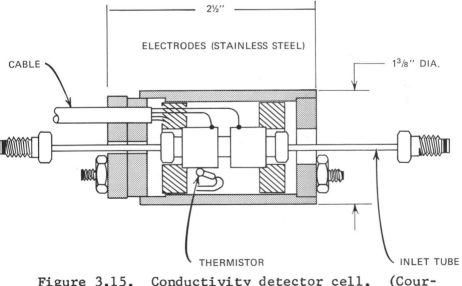

Figure 3.15. Conductivity detector cell. (Courtesy of Chromatronics, Inc.)

cell should be carefully controlled, since the conductance can change as much as 2% of the total solution conductance per degree centigrade. The thermistor in the pictured unit affords some degree of compensation for temperature. Conductivity detectors are made by Nester-Faust, Chromatronics, and LDC. The conductivity detector is most useful with aqueous mobile phases and ionic solutes. With these detectors, salt concentrations of 0.01 ppm can be detected in water.

2. Dielectric Constant

Capacitance may be used as a parameter to determine peaks eluting from a column effluent. The dielectric constant is measured at high frequency between parallel plates or concentric tubes in a small-volume cell. A standard capacitance bridge or resonant circuits can be used to make the measurement. Since dielectric constant is dependent on the polarity of the solution, solute and mobile phase should differ as much as possible. A reference cell can be used to reduce temperature effects. Cells should be extremely stable dimensionally, since small changes in the electrode gap can cause large changes in the measurement.

3. Other Devices

Detectors which have been reported or are still in their initial development stages include those using density measured by electrobalances, vapor pressure, sonic velocity, viscosity, and thermal conductivity.

There are, of course, other bulk properties which can be measured. All detectors of these types are temperature sensitive, however, and most do not have the sensitivity needed for high-speed chromatography.

References

1. General Electric Information Service Department, Book No. 802210, Program No. SIXCR$.
2. G. Deininger, A. Koneisen, and I. Halász, Chromatographia, $\underline{3}$, 329 (1970).
3. H. Felton, J. Chromatog. Sci., $\underline{7}$, 13 (1969).

4. E. Schram, in "Current Status of Liquid Scintilla-
    tion," E. D. Bransome, Jr., ed., Grune and Stratton,
    New York, 1970.
5. J. A. Hunt, Anal. Biochem., 23, 289 (1968).
6. J. G. Koen, J. F. K. Huber, H. Poppe, and G. Den
    Boef, J. Chromatog. Sci., 8, 192 (1970).
7. P. L. Joynes and R. S. Maggs, J. Chromatog. Sci., 8,
    427 (1970).
8. M. N. Monk and D. N. Raval, J. Chromatog. Sci.,
    7, 48 (1969).
9. T. B. Davenport, J. Chromatog., 42, 219 (1969).
10. R. P. W. Scott and J. G. Lawrence, J. Chromatog.
    Sci., 8, 65 (1970).
11. R. P. W. Scott, Course on Modern Practices of Liquid
    Chromatography, Wilmington, Del., April 6, 1970.
12. I. Halász, Z. anal. Chim. 234, 97 (1968).
13. R. D. Condon, Anal. Chem., 41, 4 (1969).
14. J. F. K. Huber, J. Chromatog. Sci., 7, 172 (1969).
15. M. N. Monk, J. Chromatog. Sci., 8, 491 (1970).
16. G. W. Ewing, J. Chem. Educ., 47, 9, (1970).

MANUFACTURERS MENTIONED IN TEXT

Beckman Instruments
Fullerton, California   92631

Chromatronics, Inc.
Berkeley, California   94700

E. I. du Pont de Nemours & Co.
Instrument Products Division
Wilmington, Delaware   19898

Gilford Instrument Laboratory
Oberlin, Ohio   44074

Laboratory Data Control (LDC)
Riviera Beach, Florida 33404

Nester-Faust
Newark, Delaware   19711

Nuclear-Chicago Corp.
Des Plaines, Illinois   60018

Packard Instrument Co.
Downers Grove, Illinois   60515

Phillips Electronics Instruments
Mount Vernon, New York   10550

Pye Unicam Ltd.
Cambridge, England

Varian Aerograph
Walnut Creek, California   94596

Waters Associates
61 Fountain Street
Framingham, Massachusetts   01071

CHAPTER 4

The Role of the Mobile Phase
in Liquid Chromatography

Lloyd R. Snyder

A.   INTRODUCTION

Separations by liquid chromatography depend on both the stationary phase and the mobile phase, that is, the column and the solvent. The recent development of what we have called "modern liquid chromatography" has been strongly influenced by break-throughs in column technology, and the difference between modern LC and its predecessors is largely one of columns and attendant equipment. However, the solvent also plays a critical role in maximizing the performance of LC, and the modern technique has imposed new requirements on the solvent. The selection of the right solvent system for a given LC application is therefore important.

What is required of the solvent in modern LC can be divided into two separate areas: practical aspects and the maximization of resolution. By practical aspects is meant the factors which determine whether or not an LC system is workable in a given application. The maximization of resolution refers to the achievement of adequate separation of the sample within a reasonable time. Practical requirements of the solvent include (1) column stability, (2) compatability with the detector, (3) adequate solvency for the sample, and (4) noninterference with sample recovery. In modern LC the column is normally used many times before it is discarded or repacked, because repetitive operation is convenient and also because the column often represents a considerable investment of time, effort, and expense. Consequently, the solvent system chosen for a given separation should not result in irreversible alteration of the column. Some LC detectors cannot be used with all solvent systems, particularly solvents whose composition changes with time (i.e., gradient elution). Obviously the solvent must be able to dissolve the sample, and in preparative work the solvent should not interfere with the recovery of separated sample fractions.

The effect of the solvent on sample resolution, R_S, is best discussed in terms of the resolution equation previously introduced:

$$R_S = 1/4 \frac{(\alpha - 1)}{\alpha} \sqrt{N} \left(\frac{k'}{1 + k'}\right) \qquad (4.1)$$

The resolution of two adjacent bands is determined by the separation factor, α, the column plate number, N, and the capacity factor, k'. In other words we must be concerned with separation selectivity, column efficiency, and average band migration rates. The solvent has an important effect on each of these three factors. It will suit our purpose here to assume that these three factors are independent of each other and can therefore be discussed separately. Column plate number, N, varies with solvent viscosity, while α and k' are functions of the thermodynamic properties of the solvent. The variation of average band migration rates with solvent composition depends on solvent strength: strong solvents lead to smaller k' values; weak solvents give larger k' values. Separation selectivity, α, varies with solvent composition in a more subtle fashion.

The rest of this chapter will provide a detailed discussion of the role of the solvent in LC, as outlined above. The selection of a satisfactory solvent system for a given separation can then be approached in a systematic, logical manner. In practice it is easy to eliminate the various solvents that are unsuitable on the basis of practical objections or too great a viscosity. After this preliminary screening, however, there remain a large number of common solvents from which to choose.

The selection of a solvent of the right strength from this remaining group is essential for all LC procedures except the exclusion methods (gel permeation and filtration). On the basis of the present discussion the right solvent strength can be achieved in straightforward fashion for liquid-liquid (partition) or liquid-solid (adsorption) chromatography. The control of solvent strength in ion-exchange chromatography is discussed in Chapter 8.

In some cases two or more sample bands will be incompletely resolved even after solvent strength has been optimized. For broad, multicomponent mixtures this may be due to the "general elution problem," which is discussed in Section H of this chapter. Here the problem is widely different k' values for the various components of the sample, so that with a single solvent k' is never optimum for all components. The general elution problem requires a change in solvent strength <u>during</u> separation (i.e., gradient elution), or some equivalent change in separation conditions. Although this represents an obvious complication in the selection of an overall solvent system, the principles involved are similar to those which govern the selection of a single solvent of the right strength.

Finally, we may encounter the problem of incompletely resolved bands even where k' is optimized for every pair of adjacent bands. What is needed then is either further increase in column efficiency (as discussed in Chapter 1) or a favorable change in separation selectivity (i.e., α). Alteration of α while holding k' approximately constant can be achieved by varying solvent composition. However, this is essentially a trial-and-error process, with no guarantee of success in a given case. Although some general rules can be offered to guide this empirical approach to optimum α, this aspect of solvent selection is at present the least well understood.

B. <u>PRACTICAL ASPECTS OF THE SOLVENT</u>

The importance of column stability in LC has been emphasized. Solvents should be avoided which lead to a permanent loss in column efficiency or to a change in the retention characteristics of the column. In liquid-liquid chromatography this means that the solvent and the stationary phase must be immiscible (or partially so). The prediction of immiscibility rests on an essentially empirical basis at present, although some rough guidelines will be offered in Section G. In addition to this requirement of partial immiscibility (since no two liquids are ever completely immiscible), the solvent must be presaturated with mobile phase before entering the column (see Chapters 5 and 6). In other words, the solvent and the stationary phase must be in thermodynamic equilibrium before their initial encounter within the column. In liquid-solid adsorption chromatography a similar situation is encountered, because the commonly used adsorbents are usually deactivated with adsorbed water (Chapter 6). If repeatable column operation is to be maintained, the water content of the adsorbent must not change during separation. For this reason sufficient water must be added to the solvent so that solvent and adsorbent are in thermodynamic equilibrium with respect to the activity of water in each phase. This is discussed further in Chapter 6.

Solvents which react irreversibly with the stationary phase must of course be avoided. This problem seldom arises in partition chromatography, but irreversible retention of certain solvents on chromatographic adsorbents can occasionally cause difficulty. For example, acids or traces of acids must be avoided when using basic alumina as adsorbent. Traces of impurities in the solvent can lead to alteration of the stationary phase in ion-exchange, partition, or adsorption chromatography. Because of the large volumes of solvent that flow through an LC column during its normal lifetime, stringent demands are placed upon solvent purity.

In procedures which use nonrigid gels as column packings (ion-exchange and exclusion chromatography) the gel is normally swollen to some extent by the solvent. Solvents which lead to shrinkage of the gel usually ruin the efficiency of the column. Consequently such solvents must be excluded from consideration, as discussed in later chapters.

Some attention to the compatibility of solvent and detector is usually required. Ultraviolet (UV)

detectors cannot be used with solvents that are opaque
at the wavelengths of interest. The addition of inorganic salts to the solvent can cause problems in the use
of mechanical-transfer detectors, such as the traveling
wire detector. Refractometers give reduced or zero response when the refractive index of the solvent is similar to that of one or more of the sample components.
Normally, gradient eultion cannot be used with refractometer or micro-adsorption detectors because the base
line is badly affected by changes in solvent composition.
In the case of refractometers it is possible to balance
the refractive indices of different solvent components
so as to hold the refractive index constant for the
changing solvent system (1). However, this is troublesome at best, and it leads to a substantial loss in
detector sensitivity because of increased base line
drift. Gradient elution with UV detectors also requires
some attention to the absorptivities of the various solvent components, but this is a much less serious practical problem (see, e.g., ref. 2).

Occasionally the solubility of the sample will preclude the use of certain solvents. This problem arises
mainly in preparative separations. Solvent strength and
sample solubility often vary in the same fashion as solvent composition is changed, so that it is sometimes
difficult to find a solvent that will simultaneously
provide good solvency and the right strength. Occasionally it is useful to deliberately select a solvent
that is slightly too strong and that provides poor separation of the first-eluted sample components. This
initial fraction will often be soluble in a weaker solvent, which can then be used to reseparate just the
initial fraction.

In preparative separations it is obvious that the
solvent should not interfere with sample recovery. This
is seldom a real problem, since low-boiling solvents are
desirable for other reasons (i.e., lower viscosities),
and these can usually be removed by evaporation. Reactive solvents which can polymerize to high-boiling
products should of course be avoided. For example,

acetone condenses to diacetone alcohol and higher polymers on basic alumina and should therefore not be used
with this adsorbent.

C.   SOLVENT VISCOSITY AND COLUMN EFFICIENCY

The column plate number, N, increases for small values of the plate height, H, and large values of the
column permeability constant, K'. An increase in solvent
viscosity adversely affects each of these variables. A
viscous solvent increases plate height by reducing sample
diffusion coefficients and slowing down mass transfer.
Similarly, a viscous solvent reduces column permeability
and requires a greater column pressure for a given solvent flow rate. One study (3) has shown that H in liquid-
solid chromatography increases approximately in proportion to the square root of solvent viscosity. The
constant K' is directly proportional to solvent viscosity. Combining these two factors leads to roughly a
twofold loss in N for each 2.5-fold increase in viscosity
(4), assuming constant separation times and otherwise
optimized conditions at maximum column pressures. For
constant separation efficiency (other factors equal) a
doubling of solvent viscosity must be paid for by a
doubling of separation time.

With the exception of ion-exchange chromatography, in
which aqueous solutions are commonly used, solvent viscosity can usually be kept below 0.4-0.5 centipoise with
little difficulty. A number of solvents can be found
with viscosities of 0.2-0.3 cP, and blends of these with
somewhat more viscous solvents provide a wide range of
low-viscosity mixtures from which to choose. In the case
of major solvent components comprising more than 50% of
the solvent mixture, it is seldom necessary to consider
solvents with viscosities above 0.6 cP (i.e., boiling
above about 80°).

Solvents with very low viscosities, such as pentane
or ethyl ether (0.2 cP), have often been avoided in LC
because of their very low boiling points. Low-boiling
solvents tend to form bubbles in the column and the
detector. This problem can be avoided, however, by

using a flow restrictor past the detector (e.g., a short length of small-diameter tubing). For a summary of the viscosities of some chromatographically useful solvents, see Table 4.2, column VII.

D. SOLVENT STRENGTH IN PARTITION CHROMATOGRAPHY

1. Optimization of Solvent Polarity

If the separation of two bands is to be optimized from the standpoint of maximum resolution per unit time, the k' values of the two bands must fall close to an optimum value, which is usually between 2 and 5. In the case of samples containing more than two components, the k' values of the various bands should cluster around this optimum range (e.g., $1 \leq k' \leq 10$). Most often the control of sample k' values in LC is achieved by varying solvent composition. In some cases it is necessary to vary the stationary phase as well, but this is less common. The general approach to selecting a solvent of the right strength (one which gives suitable k' values) is as follows.

From our knowledge of the starting sample we first try a solvent that we think is about right. If we have no reason for a particular initial guess, we try a solvent of intermediate strength. On the basis of the performance of the first solvent (assuming it is either too strong or too weak), we then try another solvent. Thus, if the various sample components migrate too quickly (k' values too small) with our first solvent, we consult some list of relative solvent strengths and choose a weaker solvent as our second choice. If sample k' values are too large with the second solvent, we select an intermediate solvent from our list. Such a procedure requires some ordering of the relative strengths of different solvents, varying from weak to strong. Such a grouping of solvents in order of chromatographic strength is called an eluotropic series. In this section eluotropic series or relative solvent strength for partition (i.e., liquid-liquid) separations is examined. In the following section we will consider similar series for adsorption chromatography.

The trial-and-error search for the right solvent
strength as described above can be tedious, particularly
if carried out on the column. A common alternative is
the use of corresponding open-bed separations, that is,
paper (or cellulose-layer) chromatography for partition,
and thin-layer chromatography for adsorption. A solvent
of about the right strength can be quickly selected from
such open-bed studies of the sample of interest, and the
same solvent can then be tried for the column separation.
Any further (usually minor) adjustments of solvent
strength are made on the basis of column separations.

Partition chromatography is based on the equilibrium
distribution of each sample component, X, between the
two liquid phases, m (mobile) and s (stationary):

$$X_m \rightleftharpoons X_s$$

The capacity factor, k', is given as

$$k' = \left(\frac{[X_s]}{[X_m]}\right)\left(\frac{V_s}{V_m}\right) \tag{4.2}$$

that is, the ratio of sample concentrations in each phase
times the ratio of the volumes of each phase within the
column. The volume of stationary phase can be varied
within the column (by changing the stationary-phase loading on the support) as one means of controlling k'.
However, this is relatively unimportant, and we will not
consider this possibility further. The primary aim is
to achieve the right concentration ratio by varying the
composition of the solvent m.

The relative distribution of X between the two phases
is determined by the intermolecular interactions of a
molecule of X with molecules of phases m and s. When
these interactions are stronger in one phase than the
other, the concentration of X in the corresponding phase
will be greater. What determines the relative strengths
of these sample-solvent interactions? In a fundamental
sense, the kinds and magnitudes of different intermolecular forces are responsible; more will be said about this
shortly. More generally, we can talk of the so-called

<u>polarity</u> of sample and solvent as an index of the ten-
dency of these two compounds to interact strongly. By
polarity is meant just what most chemists have in mind
when they talk of polar compounds; that is, hydrocarbons
are nonpolar, alcohols are polar. Polar compounds interact with each other quite strongly in the liquid phase,
so that polar compounds have higher boiling points (for
molecules of similar size) than nonpolar compounds.
Similarly their heats of vaporization per unit volume
are higher, since these reflect the breaking of strong,
polar bonds in the liquid phase. Any lingering doubts
as to the meaning of polarity, as well as any qualifica-
tions regarding its significance in chromatographic systems, will shortly be clarified.

   Polarity determines the distribution of X between solvent and stationary phase as follows. Consider Figure
4.1, in which we have indicated an arbitrary polarity

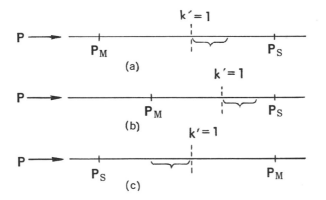

<u>Figure 4.1.</u> The interrelationship of solute polar-
ity, P, mobile-phase polarity, P_M, stationary-phase
polarity, P_S, and k' for the solute.

scale and show the polarities of the mobile phase, P_M,
and the stationary phase, P_S on that scale. In Figure

4.1a it is assumed that the mobile phase is less polar than the stationary phase, as is the usual case in partition chromatography. If a sample molecule has a polarity similar to that of the solvent ($\sim P_M$), it will prefer the mobile phase and $k' \approx 0$. Also, if the polarity of a sample molecule is similar to that of the stationary phase ($\sim P_S$), k' will be quite large ($\gg 1$). For some intermediate polarity of the sample molecule, X will distribute equally in both phases and $k' = 1$.

The bracketed range of polarities in Figure 4.1a corresponds to somewhat more polar sample molecules, those with resulting k' values in the favorable range of $1 \leq k' \leq 10$. If the components of an actual sample fall outside this range of polarity values, a change in solvent polarity is indicated. For example, assume that sample polarity (and k' value) is somewhat greater than the optimum range indicated in Figure 4.1a. In this case the polarity of the solvent must be increased. We see in Figure 4.1b that increasing the polarity of the solvent will increase the polarity of a sample molecule for which $k' = 1$. This follows because the polarity of a molecule for which $k' = 1$ will lie halfway between P_M and P_S.* As a result, the polarity of sample molecules which fall in the optimum, bracketed range will be greater in Figure 4.1b than in 4.1a. Similarly, if the system of Figure 4.1a gave k' values less than 1, the polarity of the solvent should be decreased.

In some cases involving the usual polar stationary phase plus nonpolar solvent, it is found that all possible solvents give k' values which are too small. This occurs for very nonpolar samples, such as high-molecular-weight hydrocarbons. The solution to this problem is the use of <u>reversed</u>-<u>phase</u> separation, as in Figure 4.1c, that is, a polar solvent plus a nonpolar stationary phase. In this technique the range of optimum sample polarities (the bracketed values in Figure 4.1c) is always lower than in the corresponding normal system

---

*We are assuming $V_m = V_s$, but this is not essential to the overall argument.

(e.g., Figure 4.1a versus 4.1c), because now a decrease in sample polarity means an increase in k' values.

2. Polarity Scales

How is solvent polarity measured in partition chromatography? Empirical scales of relative solvent polarities have been useful in this connection. An abbreviated listing of one such scale, that of Macek and Prochazka (5), is shown in Table 4.1. The use of such an eluotropic series is straightforward, as discussed above. For

Table 4.1

Mixotropic Series of Macek and Prochazka (5)
(Partial Listing in Order of Decreasing Polarity)

Water (most polar)	CH_2Cl_2
Formamide	$CHCl_3$
Acetonitrile	1,2-Dichloroethane
Methanol	Bromobenzene
Acetic acid	Ethyl bromide
Ethanol	Benzene
Isopropanol	Propyl chloride
Acetone	Toluene
Dioxane	Xylene
Tetrahydrofuran	CCl_4
t-Butanol	CS_2
Methyl ethyl ketone	Cyclohexane
Phenol	Hexane
n-Butanol	Heptane
n-Pentanol	Kerosine (least polar)
Ethyl acetate	
Ethyl ether	
n-Butyl acetate	
Nitromethane	
Isopropyl ether	

example, with water-methanol as the stationary phase we might try first a solvent of intermediate strength, such as CH_2Cl_2. If sample k' values prove too small (solvent too polar), benzene (less polar) might then be tried. In this way, the right solvent strength can systematically be determined.

The Hildebrand solubility parameter, δ, defines another (quantitative) scale of solvent polarities, as illustrated in Table 4.2. Values of δ range from about 6 for fluorocarbons (nonpolar) to 21 for water (polar).

Table 4.2

Solvent Properties of Chromatographic Interest

Solvent	I δ	II δ_d	III δ_o	IV δ_a	V δ_h	VI $\epsilon°$	VII η
Perfluoroalkanes[a]	6.0	6.0	0	1	0	0.25	
CFCl$_2$-CF$_3$	6.2	5.9	1.5	0	0		
Isooctane[b]	7.0	7.0	0	0	0	0.01	0.50
Diisopropyl ether	7.0	6.9	0.5	0.5	0	0.28	0.37
n-Pentane	7.1	7.1	0	0	0	0.00	0.23
CCl$_3$-CF$_3$	7.1	6.8	1.5	0.5	0		
n-Hexane	7.3	7.3	0	0	0	0.01	0.32
n-Heptane	7.4	7.4	0	0	0	0.01	
Diethyl ether	7.4	6.7	2.	2	0	0.38	0.23
Triethylamine	7.5	7.5	0	3.5	0		
Cyclopentane	8.1	8.1	0	0	0	0.05	0.47
Cyclohexane	8.2	8.2	0	0	0	0.04	1.00
Propyl chloride	8.3	7.3	3	0	0	0.30	0.35
CCl$_4$	8.6	8.6	0	0.5	0	0.18	0.97
Diethyl sulfide	8.6	8.2	2	0.5	0	0.38	0.45
Ethyl acetate	8.6	7.0	3	2	0	0.58	0.45
Propylamine	8.7	7.3	4	6.5	0.5		
Ethyl bromide	8.8	7.8	3	0	0	0.37	
m-Xylene	8.8	8.8	0	0.5	0	0.26	
Toluene	8.9	8.9	0	0.5	0	0.29	0.59
CHCl$_3$	9.1	8.1	3	0.5	0[c]	0.40	0.57
Tetrahydrofuran	9.1	7.6	4	3	0	0.45	
Methyl acetate	9.2	6.8	4.5	2	0	0.60	0.37

Table 4.2 (cont.)

Solvent	I δ	II δ_d	III δ_o	IV δ_a	V δ_h	VI $\epsilon°$	VII η
Benzene	9.2	9.2	0	0.5	0	0.32	0.65
Perchloroethylene	9.3	9.3	0	0.5	0		
Acetone	9.4	6.8	5	2.5	0	0.56	
CH_2Cl_2	9.6	6.4	5.5	0.5	0	0.42	0.44
Chlorobenzene	9.6	9.2	2	0.5	0	0.30	0.80
Anisole	9.7	9.1	2.5	2			
1,2-Dichloroethane	9.7	8.2	4	0	0	0.49	0.79
Methyl benzoate	9.8	9.2	2.5	1	0		
Dioxane	9.8	7.8	4	3	0	0.56	1.54
Methyl iodide	9.9	9.3	2	0.5	0		
Bromobenzene	9.9	9.6	1.5	0.5	0		
CS_2	10.0	10.0	0	0.5	0	0.15	0.37
Propanol	10.2	7.2	2.5	4	4	0.82	2.3
Pyridine	10.4	9.0	4	5	0	0.71	0.94
Benzonitrile	10.7	9.2	3.5	1.5	0		
Nitromethane	11.0	7.3	8	1	0	0.64	0.67
Nitrobenzene	11.1	9.5	4	0.5	0		
Ethanol	11.2	6.8	4.0	5	5	0.88	1.20
Phenol	11.4	9.5					
Dimethylformamide	11.5	7.9					
Acetonitrile	11.8	6.5	8	2.5	0	0.65	0.37
Methylene iodide	11.9	11.3	1	0.5	0		
Acetic acid	12.4	7.0				1.0^+	1.26
Dimethylsulfoxide	12.8	8.4	7.5	5	0	0.6	2.2
Methanol	12.9	6.2	5	7.5	7.5	0.95	0.60
1,3-Dicyanopropane	13.0	8.0	8	3	0		
Propylene carbonate	13.3						
Ethanolamine	13.5	8.3	Large	Large	Large		
Ethylene glycol	14.7	8.0	"	"	"		
Formamide	17.9	8.3	"	"	"		
Water	21	6.3	"	"	"		

I   Solubility parameter (calculated from the boiling point).

Table 4.2 (cont.)

II   Dispersion solubility parameter.
III   Orientation (polar) solubility parameter (approx. values).
IV   Proton-acceptor solubility parameter (approx. values).
V   Proton-donor solubility parameter (approx. values).
VI   Solvent strength parameter for adsorption chromatography on alumina (8).
VII   Viscosity (cP, 20°).
a   Average values for different compounds; the fluorochemicals (FC-75, FC-78) sold by 3M Company have similar properties and are considerably cheaper.
b   2,2,4-Trimethylpentane.
c   Other work suggests a larger value.

According to solubility parameter theory (see refs. 6,7) the distribution of a sample molecule, X, between the solvent and the stationary phase (equation 4.2) is given by the solubility parameters of X (δX), s (δs), and m (δm), and the molal volume of X ($\overline{V}X$):*

$$\log \frac{[X_s]}{[X_m]} = \overline{V}_X \frac{[(\delta X - \delta m)^2 - (\delta s - \delta X)^2]}{2.3 \ RT} \tag{4.3}$$

Here R is the gas constant and T is the absolute temperature. When the solubility parameter of X is halfway between those of the solvent and the stationary phase (i.e., $\delta X - \delta m = \delta s - \delta X$), X will be distributed equally between both phases: if $V_s = V_m$, k' = 1. Thus, the solubility parameter is a quantitative measure of what has been called polarity in the discussion of Figure 4.1. Increasing the solubility parameter of the solvent (i.e., decreasing $\delta X - \delta m$) decreases $(X)_s/(X)_m$ and k', and vice versa for decreasing δm.

---

*Equation 4.3 is strictly correct for the mole fraction ratio, rather than the concentration ratio, $[X_s]/[X_m]$. However, this need not concern us here.

Comparison of the polarity scales of Tables 4.1 and
4.2 shows a general similarity in the rankings of various solvents. However, there are some notable excep-
tions; for example, acetone (δ = 9.4) is considered
polar in Table 4.1, while CS_2 (δ = 10) is nonpolar.
These and other discrepancies in the eluotropic series
of these two tables point out a fundamental limitation
of all such series for application to partition chromatography: the relative polarity or strength of two
solvents can appear to change markedly, depending on the
type of sample. Thus any eluotropic series for partition
chromatography is only roughly constant for different
samples, and the prediction that one solvent is stronger
than another on the basis of Table 4.1 or of the δ values of Table 4.2 will not infrequently prove incorrect.
The reason for this variability of relative solvent
strength is as follows.

The concentration ratio, $[X_s]/[X_m]$, in equation 4.2
is determined by the interaction energy of the sample
molecule, X, with stationary phase s, minus the interaction energy of X with mobile phase, m. In each phase
this interaction energy is determined, not by the simple
"polarity" of each interacting molecule, but by several
different types of intermolecular forces: dispersion,
dipole induction, dipole orientation, and hydrogen bonding (see ref. 6). The overall value of δ for a given
liquid is determined by the composite of these various
interactions in the pure liquid. It is possible to
split up the Hildebrand solubility parameter, δ, into
components from these interactions, as shown in Table
4.2: δ_d for dispersion interactions δ_o for dipole interactions, and δ_a and δ_h for hydrogen-bonding interactions.
Here δ_a measures the ability of a solvent to function as
a proton acceptor, and δ_h defines the proton-donor ability of the solvent.

These various solubility parameter values (δ_d, δ_o,
etc.) are preliminary data (9) based on an approach similar to that of Hansen (10). Each of these parameters
measures a different kind of solvent "polarity." For
compounds such as nitriles or nitro compounds which have

large dipole moments, δ_o is the best single index of solvent strength or polarity. For electron donors such as ethers or amines, δ_h is a good index of solvent strength. In the case of alcohols or phenols (proton donors), δ_a measures solvent strength most accurately. The parameter δ_d is less useful as an index of solvent strength, because the dispersive interactions of different organic sample molecules usually do not differ greatly from one molecule to the next. Solvents of high δ_d will tend to exhibit a slight specificity for aromatic compounds X, and for compounds with second- or third-row (and higher functional groups (e.g., -S-, -Cl, -Br, etc.).

Although any one of the three parameters (δ_o, δ_a, δ_h) may be a preferred measure of solvent strength for a particular class of samples, in general we must consider the variation of all three--or even four--of these parameters in an eluotropic series. If the sequence of solvent strengths is to be the same (i.e., constant eluotropic series) for <u>all</u> possible samples, then δ_d, δ_o, δ_a, and δ_h must increase (or decrease) regularly for each solvent in the series. It is apparent that this is not the case for all the solvents of Tables 4.1 and 4.2, which explains their variable solvent strength toward different samples and with respect to each other. Although δ_o, δ_a, and δ_h <u>tend</u> to increase as δ increases, there are wide fluctuations for each of these parameters, as indicated in the following tabulation:

		δ	δ_d	δ_o	δ_a	δ_h
1	Triethylamine	7.5	7.5	0	3.5	0
2	Ethyl acetate	8.6	7.0	3	2	0
3	Propanol	10.2	7.2	2.5	4	4
4	Nitromethane	11.0	7.3	8	1	0
5	Methylene iodide	11.9	11.3	1	0.5	0
6	Methanol	12.9	6.2	5	7.5	7.5

Relative solvent strength series based, respectively, on δ_o, δ_a, and δ_h are different:

$$\delta_o: \quad 1 < 5 < 3 < 2 < 6 < 4$$
$$\delta_a: \quad 5 < 4 < 2 < 1 < 3 < 6$$
$$\delta_h: \quad 1 \approx 2 \approx 4 \approx 5 < 3 < 6$$

A series of several pure solvents which increase regularly with respect to each of these four parameters is difficult to obtain, particularly when we consider the other requirements on the solvent (particularly low viscosity). A simpler expedient is to consider blends of two (or more) solvents of differing polarity (e.g., n-pentane and acetone), since δ_d, δ_o, etc., vary linearly with solvent composition for solvent binaries. For example, mixtures of acetone and pentane have the following properties:

	δ_d*	δ_o	δ_a	δ_h
Pentane	7.1	0	0	0
20%v Pentane/acetone	7.0	1	0.5	0
40%v " "	7.0	2	1	0
60%v " "	6.9	3	1.5	0
80%v " "	6.9	4	2	0
Acetone	6.8	5	2.5	0

Several such series based on suitable solvent pairs can be used to cover a wide range of solvent strengths. As a corollary, if one solvent is too strong for a particular sample, and a second solvent is too weak, some blend of the two solvents will be just right. Some useful solvent pairs for covering a range in solvent strength are given in Table 4.3.

---

*The fact that δ_d decreases slightly from pentane through acetone is not expected to be important. However, a large decrease in δ_d would affect the constancy of the resulting eluotropic series for some samples, such as aromatics.

Table 4.3

Useful Solvent Pairs for Covering a Range
in Solvent Strength

Series A

  Pentane/methyl acetate (δ_d = 7.1-9.8)
  Methyl acetate/methanol (9.8-12.9)
  Methanol/water (12.9-21)

Series B

  Pentane/dioxane (δ_d = 7.1-9.8)
  Dioxane/dimethyl sulfoxide (9.8-12.8)
  Dimethyl sulfoxide/water (12.8-21)

Series C

  Pentane/50%v CH_2Cl_2-1,2-dichloroethane (δ_d = 7.1-9.6)
  50%v CH_2Cl_2-1,2-dichloroethane/nitromethane (9.6-11.0)
  Nitromethane/1,3-dicyanopropane (11.0-13.0)

E. SOLVENT STRENGTH IN ADSORPTION CHROMATOGRAPHY

Eluotropic series are found to be much more constant
(i.e., less sensitive to sample type) in adsorption chromatography than in partition chromatography. The $\epsilon°$
values of Table 4.2 define one such series for adsorption
on alumina. Weak solvents have small values of $\epsilon°$, and
strong solvents have large $\epsilon°$ values. These solvent
strength parameters vary in absolute value from one adsorbent to another (e.g., alumina versus silica), but
relative solvent strength does not change much from one
adsorbent to another (see discussion of ref. 8). In the
case of solvent mixtures, $\epsilon°$ does not vary linearly with
solvent composition as do δ, δ_d, δ_o, etc. Rather, the
strong solvent tends to concentrate into the adsorbed
phase, so that $\epsilon°$ for a mixture of two solvents, A and
B, follows the relationship illustrated in Figure 4.2.
As the concentration of the stronger solvent component,
B, increases from 0%, $\epsilon°$ rises quite steeply with increasing concentration of B (% B), and then levels off.

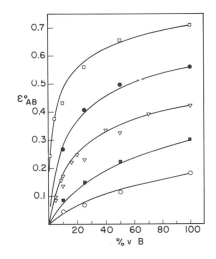

Figure 4.2. Solvent strength, $\epsilon°$, in adsorption
chromatography for binary solvents as a function
of composition (8). Alumina adsorbent; %v B refers
to concentration of strong solvent component; O,
pentane/CCl4 (B); ■, pentane/n-propyl chloride (B);
∇, pentane/CH2Cl2 (B); ●, pentane/acetone (B);
□, pentane/pyridine (B).

Adsorption chromatography $\epsilon°$ values can serve to guide
the selection of the right solvent strength, a use similar to that of δ values in partition chromatography.
   The reason for the difference in constancy of eluotropic series in adsorption and partition is as follows.
In adsorption chromatography there exists a competition
between solvent, M, and sample, X, molecules for the
adsorbent surface or stationary phase:

$$X_m + nM_a \rightleftharpoons X_a + nM_m$$

Adsorption of a sample molecule which is initially in the
solvent phase, X_m, requires the displacement of some number, n, of initially adsorbed solvent molecules, M_a,
from the adsorbent surface. The net energy of adsorption,
ΔE_a, determines the ratio of adsorbed to nonadsorbed

sample molecules (and k'); ΔE_a is given as the sum of the interaction energies for the species on the left of the above reaction, minus the interaction energies for the species on the right:

$$\Delta E_a = E_{Xa} + nE_{Mm} - E_{Xm} - nE_{Ma}$$

For a number of reasons the solution energy terms in this relationship are of lesser importance, and to a first approximation they can be ignored (see discussion of Ref. 8):

$$\Delta E_a \approx E_{Xa} - nE_{Ma}$$

That is, the net adsorption energy (which determines k') is given as the interaction energy of the sample molecule in the adsorbed phase minus n times the interaction energy of the solvent molecule in the adsorbed phase. These interaction energies are determined by the interaction of either X or M with the adsorbent surface, so that E_{Xa} is not a function of the solvent, and E_{Ma} is independent of the sample. Thus, solvent strength (i.e., the effect of the solvent on ΔE_a) is a function only of E_{Ma} and is independent of sample type; that is, $\epsilon°$ is a function only of the solvent and the adsorbent. As has been noted above, to a first approximation this is the case, particularly for weak or moderately strong solvents.

From a practical standpoint, the selection of the right solvent strength is much simpler in adsorption than in partition chromatography. Not only is it easier to find the right $\epsilon°$ value for a given sample, but also the problem of selecting immiscible stationary and mobile phases is avoided. In the case of fairly strong solvents, $\epsilon°$ values begin to show the same pronounced variability with sample type as do δ values in partition chromatography. The reasons are the same in both cases: the varying importance of dispersion, dipole, and hydrogen-bonding interactions between solvent and sample (and between adsorbent and solvent or sample in adsorption chromatography). The δ_o, δ_a, and δ_h values of Table 4.2 can be used to improve the reliability of solvent strength predictions in the same way for adsorption as

for partition. Thus solvents with large values of δ_a
will have relatively larger values of $\epsilon°$ for samples
such as alcohols and phenols. Similarly, the use of
binary solvent blends to cover a range in $\epsilon°$ values will
prove relatively more reliable as an eluotropic series.

In view of the preceding discussion, a correlation
between $\epsilon°$ and δ for different solvents is expected (11).
As shown in Figure 4.3, this is approximately the case,
so that $\epsilon°$ values for other solvents can be estimated
when their solubility parameters are known.

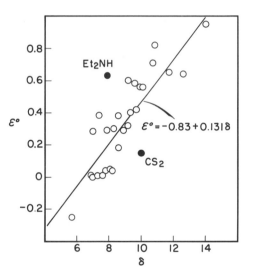

Figure 4.3. Solvent strength, $\epsilon°$, in adsorption
chromatography versus solubility parameter of the
solvent (11). Alumina adsorbent.

F.   SOLVENT SELECTIVITY IN PARTITION AND
ADSORPTION CHROMATOGRAPHY

Usually, once a solvent of the right strength for a
given sample has been found, there will be no further
problem. In some cases, however, one or more pairs of
sample bands have k' values so similar that separation
is still inadequate. In this situation the most direct

solution is usually an increase in separation efficiency, N. If the k' values of the two bands are essentially the same, however, this will not help much. Instead, we must change the relative k' values of the two compounds, while keeping k' for each band within the optimum range of $1 \leq k' \leq 10$. This can be accomplished by changing solvent composition, as discussed by Neher (12) for adsorption chromatography. In contrast to the previous remarks concerning the desirability of δ_o, δ_a, etc., varying regularly with solvent strength (so that solvents of similar strengths have similar δ_d, δ_o, etc., values), we now desire to change the relative values of δ_o, δ_a, and δ_h while keeping overall solvent strength, δ, constant. In this fashion we can take advantage of changing sample-solvent interactions to achieve differences in k' for each band.

Looking at the situation in another way, we previously wanted relative solvent strength to be the same for all sample types. But when this is the case, bands with equal k' values in one solvent will have equal k' values in all solvents. This is clearly undesirable when we are trying to separate the two bands!

As an example, assume that the two unresolved bands (partition) are a ketone, A, and an alcohol, B. Further assume that the k' values for A and B are the same with CH_2ClCH_2Cl as solvent. If we can find a solvent with approximately the same δ value as CH_2ClCH_2Cl ($\delta = 9.6$), but a substantially larger δ_a value ($\delta_a = 0.5$), the interaction of B with the solvent will be increased relative to A, so that k' for B will be reduced relative to k' for A. In this way the two bands can be separated. A likely choice for a better solvent in this specific case would be acetone ($\delta = 9.4$, $\delta_a = 2.5$), tetrahydrofuran ($\delta = 9.1$, $\delta_a = 3$), or 20% pentane/pyridine ($\delta = 9.7$, $\delta_a = 4$).

Even when the nature of two unresolved compounds is not known, the same principles can be used to maximize the chances of finding the right solvent; that is, we want to hold δ constant while varying δ_o, δ_a, and δ_h.

In this way any difference in the dipole or hydrogen-bonding interactions of the two compounds can be utilized to change their relative k' values. One example of such a group of solvents (for partition) is the following:

	δ	δ_o/δ_d	δ_a/δ_d	δ_h/δ_d
Methylene iodide	11.9	0.1	0	0
Acetonitrile	11.8	1.2	0.4	0
50%v Methanol/ethanol	12.0	0.7	0.9	0.9
25%v Propylamine/dimethyl sulfoxide	11.8	0.8	0.7	0

Variation of the solvent in order to change selectivity is somewhat more complex in adsorption than in partition chromatography, although the same approach as that described above can be used. When binary solvents are employed, it must be kept in mind that ϵ^o is no longer predictable by δ (see ref. 11). For a fuller discussion of solvent selectivity in adsorption see ref. 8.

In the case of weaker solvents, where dipole and hydrogen-bonding interactions are less important, it might be anticipated that solvent selectivity in adsorption chromatography would be less pronounced. Although this is roughly the case, quite substantial changes in sample k' values can be achieved by varying the solvent, as is illustrated in Table 4.4 for the two compounds 1-acetonaphthalene and 1,5-dinitronaphthalene, eluted from 4% H_2O/alumina. The separation factor, α, for these two compounds can be varied over a sevenfold range, while holding k' within the desired range of $1 \leq k' \leq 10$. For these and other substituted naphthalenes and solvents, no correlation between solvent selectivity and the δ_o, δ_a, or δ_h values of sample or solvent was found. Rather, solvent selectivity varies smoothly for all sample compounds from nonpolar solvents ($\delta = \delta_d$) to solvents which contain increasingly polar components (e.g., pyridine and dimethyl sulfoxide). As a result, maximum selectivity for this series of compounds on alumina occurs with solvents of the first type (nonpolar) or the last type

Table 4.4

Solvent Selectivity in Adsorption
Chromatography on Alumina (13)

Solvent (solution in pentane)	k'		
	1-Aceto-naphthalene	1,5-Dinitro-naphthalene	α
50%v Benzene	5.1	2.5	0.5
25%v Ether	2.5	2.9	0.8
23%v CH$_2$Cl$_2$	5.5	5.8	0.95[a]
4%v Ethyl acetate	2.9	5.4	1.8
5%v Pyridine	2.3	5.4	2.4
0.05%v Dimethyl sulfoxide	1.0	3.5	3.5[b]

[a] 14,000 effective plates required for separation (R_s = 1.5).

[b] 40 effective plates required for separation (R_s = 1.5).

(dilute solutions of strong solvents), with no advantage for intermediate solvent systems.

G.   PHASE IMMISCIBILITY IN PARTITION CHROMATOGRAPHY

It was noted earlier that immiscible liquids for partition chromatography are selected empirically, because immiscibility cannot be predicted with much accuracy. It is useful to look briefly at the general factors which determine immiscibility, in terms of solubility parameter theory. The heat of mixing two liquids, i and j, is predicted from theory to be

$$\Delta H^M = (X_i V_i + X_j V_j)\,(\delta_i - \delta_j)^2\,\emptyset_i \emptyset_j \qquad (4.4)$$

Here X_i and X_j are the mole fractions of i and j, V_i and V_j are their molal volumes, δ_i and δ_j are their solubility parameters, and $\emptyset_i$ and $\emptyset_j$ are their volume fractions in the final solution. According to equation 4.4, ΔH^M is always positive, a circumstance which works against phase miscibility. The larger is ΔH^M, the less likely is miscibility. Immiscibility is seen to be favored by large molal volumes of i and/or j and by a large difference in their solubility parameters. Since the viscosity of the mobile phase should be low, its molal volume cannot be very large. Since there is less reason to be concerned about the viscosity of the stationary phase, higher-molecular-weight stationary phases can be used to increase phase immiscibility.

As equation 4.4 stands, it is relatively unreliable, because δ is a composite of dispersion, dipole, etc., contributions. An improvement can be affected by replacing the term $(\delta_i - \delta_j)^2$ by

$$(\delta_d{}^i - \delta_d{}^j)^2 + (\delta_o{}^i - \delta_o{}^j)^2 - \delta_a{}^i\delta_h{}^j - \delta_h{}^i\delta_a{}^j.$$

Although this expression is not rigorous, it correctly predicts that phase immiscibility is favored when the differences between δ_d and δ_o values for the two solvents are large, and when the cross products of δ_a and δ_h are small.

H. THE GENERAL ELUTION PROBLEM

Figure 4.4 illustrates an example of the general elution problem in LC; here is shown the separation of a synthetic eleven-component mixture by adsorption chromatography on silica, with pentane as solvent. Because of the spread in k' values for the various components of this sample, initially eluting bands have k' values which are too small, and the last-eluted bands have excessively large k' values. As a result the first bands off the column are poorly resolved, while the last bands require an excessive time for elution and have broadened to the point where detection sensitivity is significantly lowered. In more extreme examples of the general elution problem, some initially eluting bands would be completely unresolved, and bands that eluted later would never be observed.

An effective solution to the general elution problem requires a change in band migration rates _during_ separation. There are several different ways whereby this can be accomplished:

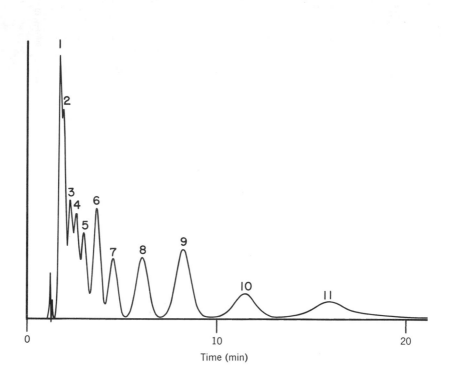

Figure 4.4. Separation of a synthetic mixture on silica, with n-pentane as solvent.

- Gradient elution or solvent programming.
- Temperature programming.
- Flow programming.
- Repeated separation under changed conditions.
- Coupled columns (stationary-phase programming).

We will discuss each of these techniques in order, pointing out its advantages and limitations and discussing the conditions for optimum separation. (For a detailed treatment of this subject matter see refs. 14-16.)

Before these individual techniques for solving the general elution problem are examined, several practical criteria which influence the selection of a particular technique should be noted. Obviously maximum resolution

per unit time is desired at all points in the chromatogram (particularly at the beginning). The technique
should be applicable to samples with a wide range of k'
values; in adsorption chromatography, it is possible to
encounter, for a given solvent, a range in sample k' values of 10^{10}. We should consider whether the technique
is to be applied to the routine, repetitive analysis of
similar samples or to the initial analysis of a completely unknown mixture. The equipment required for applying
a given technique should be as uncomplicated as possible,
and the procedure itself should be simple. It is desirable that column regeneration not be required after
separation; that is, application of the technique should
leave the column in its initial condition at the end of
separation. The technique should be usable with all
types of LC detectors. Detection sensitivity (which
varies markedly with k' in normal elution) should be uniformly high for all bands.

An overall summary of the five techniques with respect
to these various requirements is given in Table 4.5.

1. Gradient Elution

Gradient or stepwise elution involves the change of
solvent composition during separation in such a way that
solvent strength increases from the beginning to the end
of separation. In this way the k' value of each eluted
band is approximately optimum during its movement through
the column. As a result gradient elution provides unsur-
passed resolution per unit time for wide-range samples,
and this advantage over other techniques becomes more
pronounced as the ratio of k' values for the last-eluted
and the first-eluted bands increases. When this ratio
exceeds a value of about 1000, gradient elution and repeated separation are the only useful techniques.

The equipment and the experimental technique required
for high-performance LC with gradient elution tend to be
relatively complex, and this lack of simplicity has seriously limited its application. After separation by
gradient elution, the column must be regenerated by washing away the last part of the solvent gradient. With

Table 4.5

Evaluation of Five Techniques for Solving the General Elution Problem[a]

Technique	Resolution	Applicability to Wide Range of Samples	Equipment Complexity	Need for Column Regeneration	Inapplicable with Any LC Detectors?	Detection Sensitivity for Last Peaks	Special Problems
Gradient elution	++	++	0	+	Refractometer, heat-of-adsorption	++	
Temperature programming	0	0	+	+		++	
Flow programming	0	0	++	+	Refractometer, heat-of-adsorption	0	
Repeated separation	+	++	+	0		+	Need for composited analysis means multiple-sample handling for each sample.
Coupled columns	+	+	++	++		+	

a ++ is quite good, + intermediate, 0 not so good.

gradient elution, detection sensitivity is maximized for late-eluting bands, but the technique cannot be used with refractometer or heat-of-adsorption detectors.

Despite these limitations, however, the outstanding performance of gradient elution for broad-range samples makes it the technique of choice in this case. Similarly, gradient elution is well suited to the analysis of samples of unknown complexity, since good resolution is automatically provided for a wide range of sample polarities. For a detailed discussion of the experimental optimization of gradient elution, see ref. 15. The requirements in regard to the solvent gradient or program are similar to those for an eluotropic series; δ or ϵ° should increase linearly with time, and δ_0, δ_a, and δ_h should increase regularly throughout separation (i.e., not fluctuate back and forth).

2. Temperature and Flow Programming

Temperature programming normally consists of an increase in column temperature during separation (see, e.g., ref. 17). As a result the k' values of different bands are lowered during separation, just as in gradient elution. The major problem in temperature programming is reduced resolution, as a result of changes in solvent viscosity which accompany the variation in solvent temperature. Temperature programming is seldom used in LC because of reduced resolution and the limited ability of the technique to improve the separation of broad-range samples. In adsorption chromatography, column regeneration is normally required after separation, because it is impossible to maintain the adsorbent water content constant during temperature programming (see Chapter 6).

Flow programming consists of a reduction in solvent velocity during the initial stage of the separation, and an increase in solvent velocity during the last part. In this way resolution is effectively transferred from the latter part of the chromatogram (where often it is not needed) to the front part (where it is needed in the general elution problem). Flow programming provides

only a marginal increase in front-end resolution, and the
technique is useless for samples of very broad range.
Its primary advantage is simplicity of equipment and experimental technique, which will occasionally make its
application worthwhile.

3. Repeated Separations

Repeated separation refers to the separation of the
same sample two or more times, using different experimental conditions in each separation. In this way k'
can be optimized for different groups of sample components, and a final analysis can be obtained by composit-
ing the results of each separation. An example of this
application is the well-known two-column procedure for
amino acid analysis. Acidic amino acids are separated
and determined on one column, while basic amino acids
are determined on another. The disadvantages of repeated separation include the need for column regeneration when all of the sample components are not eluted
during the separation (e.g., in the determination of the
acidic amino acids mentioned above), plus the complexity
of two or more separations (versus one separation by
other techniques). Repeated separation is also somewhat
wasteful of the sample, and this factor would be significant in preparative separations.

4. Coupled Columns

The term coupled columns refers to a new technique
for solving the general elution problem (16). One experimental arrangement for this technique is illustrated
in Figure 4.5. The sample is initially separated into
rough fractions of different polarity on the fore-column
(1). The initially eluting, less polar fraction is
diverted to column 2, the intermediate fraction to column 3, and the last, most polar fraction to column 4.
This is accomplished by varying the setting of the four-
way valve between column 1 and columns 2-4. By varying
the stationary phase or its relative amounts in columns
2-4, sample k' values are made to be greatest in column

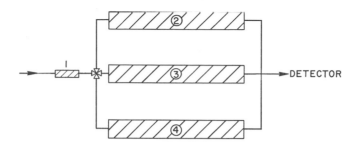

Figure 4.5. Column arrangement for separations by coupled columns. Columns 1 and 4, weak retention; column 3, intermediate retention; column 2, strong retention.

2, less in column 3, and least in column 4. This provides good resolution and rapid elution of the various fractions sent to the individual columns. The resulting overall resolution per unit time is not much inferior to that provided by gradient elution, and samples of fairly broad range can be handled conveniently.

The equipment and the experimental procedure involved in coupled-column operation are extremely simple, there is no need for column regeneration (the same solvent flows through all columns), any LC detector can be used, and detection sensitivity for the last-eluted bands is almost as good as in gradient elution. For these reasons, the coupled-column technique should soon be widely applied as a solution for the general elution problem. A variety of column packings are available for ready application to different types of separation problems. Pellicular (e.g., Zipax® or Corasil®) and porous (high-surface-area) packings can be used to vary k' values by varying column capacity (surface area in adsorption, stationary-phase loading in partition, charge concentration on the resin in ion-exchange). Other column packings can readily be conceived for the same purpose (see, e.g., ref. 16).

Figure 4.6 compares the application of gradient elution (b), coupled columns (c), and flow programming (d) to normal elution separation, shown in Figure 4.4.

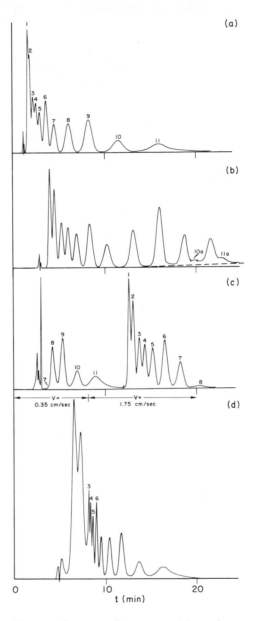

Figure 4.6. Comparison of separation by normal elution (a), gradient elution (b), coupled columns (c), and flow programming (d) (16).

References

1. K. J. Bombaugh, R. N. King, and A. J. Cohen, J. Chromatog., 43, 332 (1969).
2. L. R. Snyder and D. L. Saunders, J. Chromatog. Sci., 7, 195 (1969).
3. L. R. Snyder, Anal. Chem., 39, 698 (1967).
4. L. R. Snyder, J. Chromatog. Sci., 7, 352 (1969).
5. I. M. Hais and K. Macek, "Paper Chromatography," Academic, New York, 1963, p. 115.
6. J. H. Hildebrand and R. L. Scott, "The Solubility of Nonelectrolytes," 3rd ed., Dover, New York, 1964.
7. J. H. Hildebrand and R. L. Scott, "Regular Solutions," Prentice-Hall, Englewood Cliffs, N.J., 1962.
8. L. R. Snyder, "Principles of Adsorption Chromatography," Marcel Dekker, New York, 1968.
9. B. L. Karger, R. A. Keller, and L. R. Snyder, unpublished data.
10. C. M. Hansen, Ind. Eng. Chem. Prod. Res. and Develop., 8, 2 (1969).
11. R. A. Keller and L. R. Snyder, J. Chromatog. Sci., to be published.
12. R. Neher, in "Thin-Layer Chromatography," G. B. Marini-Bettolo, ed., Elsevier, Amsterdam, 1964, p. 75.
13. L. R. Snyder, unpublished data.
14. L. R. Snyder, Chromatog. Rev., 7, 1 (1965).
15. L. R. Snyder and D. L. Saunders, J. Chromatog. Sci., 7, 195 (1969).
16. L. R. Snyder, J. Chromatog. Sci., 8, 692 (1970).
17. R. P. W. Scott and J. G. Lawrence, J. Chromatog. Sci., 8 (1970).

PART TWO
THE PRACTICE OF LIQUID CHROMATOGRAPHY

CHAPTER 5

The Practice of Liquid-Liquid Chromatography

Joseph J. Kirkland

A.   INTRODUCTION

Liquid-liquid chromatography, sometimes called liquid-partition chromatography, has been a powerful technique for high-resolution separations since its inception in 1941 by Martin and Synge (1). This technique has been applied less often for analytical purposes, however, than the newer methods of gas chromatography or thin-layer chromatography. Recently, improvements in column technology, along with the development of better apparatus, have led to renewed interest in this technique. Theoretical insight, specialized column packings, sensitive detectors, and reproducible pumping systems have all been combined to make high-pressure, high-speed, liquid-liquid chromatography a practical means of separation.

Much of the renewed interest in liquid-liquid chromatography has been manifested by workers who are skilled in gas-liquid chromatography; therefore, modern LLC has taken the form of GLC. For example, LLC systems have been described which have column efficiencies and analysis times comparable to gas chromatographic approaches. Therefore, modern LLC is analogous to GLC.

The LLC column consists of a bed of finely divided solid (the support), usually inert, on which a stationary partitioning phase is fixed. The mobile phase flowing through the column is thus in contact with the stationary phase over a very large interface. By this process, equilibrium distribution of the solute (sample) between these two phases rapidly takes place. Separation of the different components of a mixture results from the differing distributions of the various solutes in the unlike phases. The present discussion will deal only with elution chromatography, whereby the sample components migrate down the column with different velocities (resulting from differing interactions within the

column) so that the separation improves as the column
length is increased.

The general theory of liquid chromatography has already been described in Chapter 1. In addition, detailed
theoretical discussions are available elsewhere (2-5).
The objective of this chapter is to discuss the state of
the art in the application of LLC to rapid, high-
efficiency separations.

B. ESSENTIAL FEATURES OF LIQUID-LIQUID CHROMATOGRAPHY

Some general comments regarding the essential features
of liquid-liquid chromatography, as compared to other
chromatographic approaches, should be useful in ensuring
the proper application of this technique to separation
problems.

Versatility for Sample Types. When compared to other
forms of chromatography, including other techniques of
liquid chromatography, LLC remains one of the more versatile procedures. This versatility is based largely on
the selection of partitioning phases which may be used to
accomplish the desired separation. Unique chemical interactions may be used to perform certain separations
which might be difficult by other chromatographic techniques. Liquid-liquid chromatography can be applied to
a wide variety of sample types, both polar and nonpolar.
In the most widely used form, the stationary phase is a
polar material, while the mobile phase is considerably
less polar. This arrangement is used for separating the
more strongly polar compounds, these materials being
preferentially retained in polar stationary phases. To
separate nonpolar molecules, the phases are reversed.
This technique is sometimes called "reversed-phase"
liquid-liquid chromatography.

Versatility of Retention System. There is no "universal partitioning system" for all solutes such as that
exemplified by silica gel in LSC. Theoretically, however, there is an almost infinite capability for separation by selecting the appropriate pairs of partitioning
liquids. Limitations will be discussed later.

Reproducibility of Packings. One of the strong practical advantages of LLC is the ability to produce columns
which can reproducibly carry out desired separations.
This is in contrast to columns made with certain adsorbents; the method of preparation of the adsorbent is
often less reproducible from batch to batch than would
be desired.

Speed of Separation. Recent progress in LLC technology now permits separations as fast as any obtainable
with the other forms of chromatography. Although LLC
column performances of 80 theoretical plates (or 12
effective plates) per second have been reported for moderately retained solutes (6), the efficiency of liquid
chromatographic columns generally remains lower than that
of gas chromatography (see Chapter 1). This simple comparison of column efficiency between gas and liquid chromatography is not the entire story, however. In many
cases, it is possible to achieve faster separation of a
given pair of solutes by liquid-partition chromatography
because of the greater possibility of obtaining large α
values by using selective interactions.

Solute Stability. Liquid-liquid chromatographic separations may be employed with a wide range of labile or
reactive solutes with excellent results, since this technique may be carried out at ambient or less than ambient
temperatures, using materials which are devoid of chemical reactivity. When properly engineered, this technique
exhibits none of the problems with thermal degradation
that often occur in the gas chromatographic separation
of many polyfunctional compounds. Catalytic effects
which sometimes influence isomerization, hydrolysis, etc.,
during LSC separations also are usually not found.

Quantitative Features. Liquid-liquid chromatography
is particularly applicable to the more difficult quantitative techniques which involve high-precision assays
and trace analyses at the less than parts-per-million
level. A combination of some of the advantages discussed
above makes LLC one of the most precise of the liquid
chromatographic techniques.

C.   EXPERIMENTAL FACTORS AFFECTING COLUMN EFFICIENCY

1. Supports for Liquid-Liquid Chromatography

Two general types of supports are used for liquid-liquid chromatography; porous supports and superficially porous (thin-layer or pellicular) supports. Porous supports include silica gel, diatomaceous earth materials (e.g., Chromosorb®), and porous silica beads, such as Porasil®. These supports contain pores throughout the structure and have high surface area. The superficially porous, or porous thin-layer supports, which are composed of particles with an impervious center and a thin, porous shell, include Zipax® controlled surface porosity support, Corasil®, and surface-etched beads. A summary of the general types of materials is shown in Table 5.1.

Table 5.1

Some Supports for Liquid-Liquid Chromatography

Name or Type	Supplier[a]
Porous materials	
Diatomaceous earths	Analabs, Applied Science
Porous silica beads (Porasil®)	Waters Associates
Silica gel	Davidson Chemical
Superficially porous materials	
Zipax®	E. I. du Pont de Nemours & Co.
Corasil®-I	Waters Associates
Surface-etched beads	Corning Glass Works[b]

[a]Other suppliers available for some materials.

[b]Not commercially available at this writing.

Porous Supports. The advantages of this class of supports are that they are less expensive, are relatively available, and possess greater capacity for solutes because of their high surface area. Their main disadvantage is that both the mobile and the stationary phases can be trapped in deep stagnant pools within the totally porous structure. This situation results in a significant band broadening due to resistance to mass transfer from the diffusional processes in the stagnant liquid (see Chapter 1 for a more detailed discussion). Silica gel and the diatomaceous earths are also irregular in shape, making it more difficult to prepare column beds with uniform structure. Dry-packing techniques, particularly, are more difficult with the small particles of these irregular materials. The total porosity of the porous supports, coupled with their irregular particle shapes, leads to lower column efficiencies when used in LC separations at high carrier velocities.

Superficially Porous Supports. Some of the advantages of the porous thin-layer or superficially porous supports were indicated in Chapter 1. These dense, spherical materials are readily and reproducibly packed into columns with high homogeneity. High column efficiencies are possible with these supports, even with high carrier velocities. Such rigid materials can also be used at high pressures without deformation. The disadvantages associated with these supports are that they are more costly and have lower sample capacity.

A schematic of a cross section of a particular porous layer material, the controlled surface porosity particle (7,8) is shown on the left of Figure 5.1. This support consists of spherical siliceous particles with a porous surface of controlled thickness and pore size. An electron photomicrograph of a cross section of a controlled surface porosity particle is shown on the right side of Figure 5.1. When using a support of this type, one disperses the stationary phase (or the sorbent) throughout the surface of the porous shell.

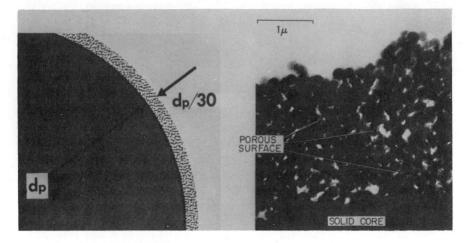

Figure 5.1. Cross sections of controlled surface porosity support particles. Left: Schematic. Right: Electron photomicrograph. Particle size - about 30 μ; average pore size - approximately 1000 Å. (Reproduced from refs. 12 and 26 by permission of publisher.)

The stereoscanning electron micrograph of the Zipax® controlled surface porosity support in Figure 5.2 shows the spherical nature of the particles and the regularity of the porous surface. Normally, the entire pore structure of the particle is not completely filled with the partitioning phase; rather, sufficient material usually is added so as to form a film about 100-400 Å in average thickness throughout the porous structure. The pores within the outer shell of the Zipax® are relatively large, allowing most solute molecules ready access to the stationary phase contained within the porous structure.

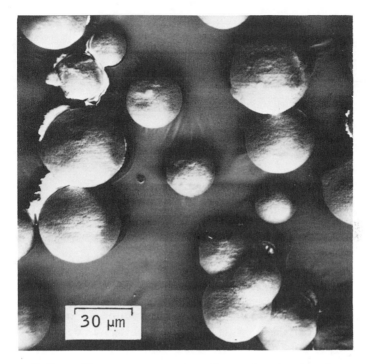

Figure 5.2. Stereoscan scanning electron micrograph
of controlled surface porosity support particles.
(Reproduced by courtesy of Ronald E. Majors and M. M.
Sieminski, Celanese Research Co., Summit, N.J.)

The general chromatographic characteristics of each of
the porous thin-layer or superficially porous supports
show similarities. A comparison of the relative efficiencies of some of the supports mentioned in previous
chapters is shown in Figure 5.3. Since data obtained
from different particle sizes of the various supports are
compared, reduced plate heights, h, are plotted against
reduced carrier velocities. (This comparison is not
exact and is actually biased in favor of supports with
larger particles, as discussed in Chapter 9.) The amount
of liquid on the various supports was adjusted so that
the stationary liquid film thickness was approximately

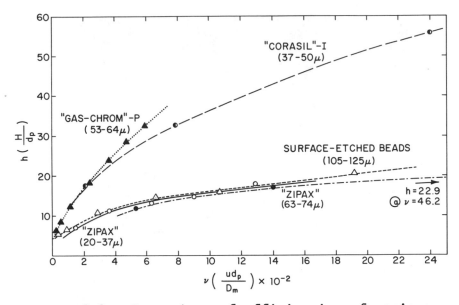

Figure 5.3. Comparison of efficiencies of various supports. Columns = 1 meter, 2.1 mm i.d.; carrier = hexane; sample = 1 μl of 5 mg/ml benzyl alcohol in hexane; temperature = 27°C; stationary phase = β,β'-oxydipropionitrile, 0.25% on surface-etched beads, 1% on Zipax® and Corasil®, 4% on Gas-Chrom®-P; UV detector.

the same in all supports, taking into account the variances in surface area and density.

2. Support Particle Size

As indicated by the discussions in Chapter 1, support particle size has a major effect on column efficiency. Irregular supports, such as the diatomaceous earths and silica gel, are difficult to dry-pack into homogeneous columns at particle sizes of less than about 50-60 μ (9). Dense, spherical supports, such as Zipax® and glass beads, can be dry-packed with particle sizes down to about 20 μ, with good results. It may be noted in

Figure 5.3 that 20-37 μ and 60-74 μ fractions of Zipax®
produce essentially the same plots of reduced plate
height versus reduced carrier velocity, indicating that
it is possible to obtain by dry-packing techniques an
efficiency for the smaller Zipax® particles equivalent
to that for the larger particles. With particles below
20 μ, the plate heights of glass bead (10) and Zipax®
(11) columns produced by dry-packing techniques actually
are greater. Slurry-packing techniques, to be discussed
later, may be used to pack irregular particles of < 50 μ
and spherical, superficially porous particles of < 37 μ.

3.   Partitioning-Phase Loadings

Except at very high stationary-phase loading, column
efficiency generally is not greatly affected by partitioning film thickness. Most high-speed liquid chromatographic columns are actually mobile-phase mass transfer
limiting. Therefore, if the LLC column has been properly designed in terms of the type and deposition of
stationary phase, its efficiency is not significantly
affected throughout a rather broad range of liquid loading. However, there are some secondary factors which do
affect the efficiency of columns, so that plate height
increases as the concentration of the partitioning
medium exceeds a certain level.

This situation is illustrated in Figure 5.4, which
shows the effect of the level of β,β'-oxydipropionitrile
on columns prepared with an experimental controlled surface porosity support (12). The plate heights, H, for
two solutes (constant carrier velocity) are approximately equal up to liquid-phase concentrations of about
0.75%. Above this level, the plate height increases.
Apparently, when a certain minimum amount of liquid is
placed on the particles, a liquid film forms on the outside of the porous shell. When this film is sufficiently thick, the attraction of the outside liquid films
between individual particles causes some aggregation,
making it more difficult to pack a homogeneous bed by
dry-packing techniques.

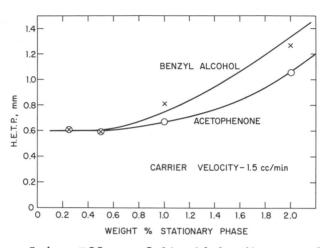

<u>Figure 5.4</u>.   Effect of liquid loading on plate
height.   Column = 0.5 meter, 3.2 mm i.d.; 20-37 μ
controlled surface porosity support; carrier =
hexane; temperature = 27°C; sample = 5 μl of a
hexane solution containing 1 mg/ml acetophenone
and 5 mg/ml benzyl alcohol; UV detector.

It should be pointed out, however, that although optimum column efficiency is obtained on Zipax® with liquid
loadings up to about 0.75%, it is often desirable to ac-
cept a modest loss in column efficiency and to increase
the loading to as much as about 1.5% stationary liquid.
The slight loss in column efficiency resulting from the
heavier loading is offset by the increased solute partition ratios, increased sample capacity, and improved column stability.   The optimum level of partitioning-phase
loading for a particular support is largely attendant on
factors such as the viscosity of the stationary phase,
the molecular weight, and the type of liquid (polarity,
etc.).   A good rule of thumb is that one should use the
highest liquid-phase loading which does not grossly degrade the efficiency of the column packing.   Suggested
liquid-phase loadings for some of the LLC supports are
shown in Table 5.2.

Table 5.2

Suggested Liquid-Phase Loadings for Various
Liquid Chromatographic Supports

Material	Particle Size, μ	Approx. Surface Area, m^2/g	Maximum Liquid Loading, wt %	Optimum Liquid Loading, wt %
Diatomaceous earth (Gas Chrom®-P, Chromosorb®, etc.)	53-68	1	30	5-10
Zipax®	< 37	1	2.0	0.5-1.5
Corasil®-I	37-50	7	1.5	0.5-1.0
Surface-etched beads	37-44	0.4	0.5	0.2-0.4

On the basis of the theory presented in Chapter 1, a straight-line relationship should exist between adjusted retention volume (solute retention volume less volume for unretained peak) and per cent liquid phase. This relationship is verified by the plots shown in Figure 5.5. (These data also indicate that the solvent evaporation method of coating the controlled surface porosity support with the stationary liquid phase is precise.) The extrapolated plots in Figure 5.5 show some retention at zero liquid phase, indicating that uncoated Zipax® controlled surface porosity support is a weak adsorbent. This support is essentially inert when coated with a polar liquid phase in concentrations useful for LLC.

As a consequence of the relatively high specific surface area required in the thin porous layer surrounding the impervious core of superficially porous supports, these materials display activity as adsorbents. As mentioned above, Zipax® (nitrogen surface area, ca. 1 m^2/g) is a weak adsorbent. However, Corasil®-I (nitrogen surface area, ca. 7 m^2/g) apparently is an active adsorbent capable of liquid-solid separations without modification (13). This adsorbent activity is manifested in several ways. For example, relative retention is a function of

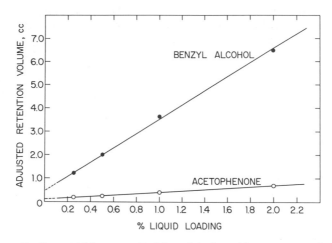

Figure 5.5. Effect of liquid loading on solute retention volume. Conditions same as in Figure 5.4;
carrier = 0.45 cc/min. (Reproduced from ref. 12
by permission of publisher.)

the amount of stationary phase on the support (14). In
addition, dry support will irreversibly adsorb stationary phase when conditioned in a column with carrier
saturated with the stationary phase. From β,β'-oxydi-
propionitrile-saturated heptane, Zipax® adsorbs about
0.25% (w/w) of the stationary phase, Corasil®-I about
1.1%, at the steady state. A simple way to coat
Corasil®-I reproducibly is to fill the column with dry
support and to flow stationary-phase-saturated carrier
through overnight (14).

4. Column Internal Diameter

Although column efficiency is dependent on the particle size of the support and the length of the column
(15), the plate height is also significantly influenced
by the internal diameter of the column. For analysis,
columns with 2-3 mm i.d. provide a good compromise between sample capacity, amount of packing used, amount of
solvent required, and column efficiency. Dry-packed
columns below about 2 mm i.d. have been found to be less

efficient (12). On the other hand, highly efficient columns of 10-11 mm i.d. have been demonstrated (16,17).
Peak-shape anomalies have been observed in columns with
internal diameters of 5-7 mm, depending on column length
(11,17). This distortion probably can be attributed to
wall effects, in which solute molecules that have reached
the wall where the carrier velocity is greater than average are moved more quickly down the column than are molecules in the center, where the carrier velocity is less
than average. In narrower-bore columns, the solute transfers in and out of these wall areas sufficiently so that
the solute band eluting from the column is essentially
symmetrical. When the internal diameter of the column
is such that the solute molecules transfer to the walls
but do not have enough time to move back into the center
of the column before they are eluted, trans-column
equilibration of the solute band does not occur. The
zone then elutes unsymmetrically, sometimes with distorted trailing edges or occasionally with doublet peaks.
   Highest column efficiency apparently results when the
solute molecules never reach the disturbed wall area, a
condition called the "infinite-diameter" column phenomenon (15). The internal diameter of an "infinite-
diameter" column may be calculated by the equation

$$d_c = (2.4d_pL)^{-2}$$

For instance, a column 0.5 m long (L) packed with 30 μ
particles (d_p) must have a minimum internal diameter of
6 mm (d_c) to be an "infinite-diameter" column.

5. Sampling

   When working with liquid-liquid chromatographic columns, it is desirable to introduce a sample solution into
the column which has been prepared in the exact carrier
composition. If sample solutions prepared in a solvent
which has a higher solubility for the stationary phase
are employed, a small amount of the stationary phase is
dissolved by the solvent in the sample aliquot with each
introduction. This dissolution may be eliminated by
first saturating the solvent used in preparing the

sample with the stationary phase; otherwise, "stripping"
of the stationary phase from the column support can
occur. Columns containing packings with chemically
bonded phases (discussed in Section F.2 of this chapter)
do not require this precaution, and samples usually may
be prepared in whatever solvent is convenient for the
system. Polar solvents used to prepare solutions of
samples do not cause changes in columns with bonded-phase
packings but may generate a large initial peak in the
chromatogram, particularly when detectors depending on
bulk properties are used.

An aliquot containing a relatively insoluble sample
dissolved in a polar solvent must not be injected into
a nonpolar carrier, lest the sample precipitate in the
carrier stream. This situation can result in catastroph-
ic peak broadening and tailing.

The manner in which a sample is injected into a column
and the volume of sample aliquot both have important
effects on column efficiency. Highest column efficiencies are sometimes obtained by injecting the sample
directly into the top and center of the column packing.
In this manner, diffusion of the solute bands is held to
a minimum during passage down the column. Introduction
of the sample into the solvent stream above the column
packing generally results in the solute being swept into
the top of the column as a diffuse band, with resulting
decrease in column efficiency for sample components,
particularly those which are rapidly eluted. Properly
engineered microvalves can also be used for efficient
sampling, as described in Chapter 2. Peak symmetry,
important in obtaining maximum resolution and precise
quantitative measurements, can be influenced greatly by
the method of introduction.

To maintain optimum column efficiency, it is desirable not to exceed a certain maximum sample volume with
a column of a particular internal diameter. Figure 5.6
shows the effect of sample volume (constant solute
weight) on the plate height of a 1 m x 2.1 mm i.d. LLC
column with controlled surface porosity support. These

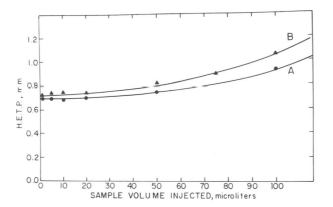

<u>Figure 5.6.</u> Effect of sample volume on column effi-
ciency. Column = 1 meter, 2.1 mm i.d., rest same
as in Figure 5.7. Curve A, column 1 week old;
curve B, column 21 days old. (Reproduced from ref.
<u>26</u>, by permission of publisher.)

data indicate that sample volumes up to about 50 μl can
be injected into this particular column without significantly affecting column efficiency. Larger sample volumes are tolerated if sample k' values are large. The
effect of sample volume on the efficiency of columns of
other internal diameters is generally a linear function
of the column cross-sectional area of the same column
material, and is dependent on the support, stationary-
phase loading, temperature, and type of liquid-liquid
partitioning system.

Column efficiency is also affected by the amount of
total sample placed on the column, for overloading of the
column can drastically decrease efficiency. For analytical studies, it is usually desirable to use the least
amount of sample commensurate with the means of detection
available. Many analyses are carried out with 1-25 μg
of material with narrow-bore columns. Even with LLC columns prepared with porous thin-layer supports, a fractional milligram of material can often be chromatographed
without serious peak broadening due to overloading.
Although columns with these supports are not recommended

for preparative chromatography, it is feasible and convenient to be able to isolate unknown substances for identification by using high-efficiency analytical colums of these materials.

D. COLUMN PREPARATION

Since the column is the heart of a liquid chromatographic separation, it is important that maximum efficiency be built into the column in its preparation. The construction of columns is generally much more critical for liquid than for gas chromatography. Imperfections in the packing of LC columns results in much larger relative peak broadening (see Chapter 1 for discussion). It is critical, therefore, that the partitioning phase be placed on the support in a thin, homogeneous film, and that the support be uniformly packed into the column. There are several methods by which these two operations may be satisfactorily performed, although it should be recognized that any of these procedures, in light of future developments, may not be optimum.

1. Preparation of Liquid-Coated Supports

Solvent Evaporation Technique. A generally useful technique for coating supports with a liquid stationary phase is the solvent evaporation procedure widely used in GC. The column packing is prepared by slurrying the support with a solution of the stationary liquid phase in a volatile solvent such as dichloromethane. The volatile solvent is removed by evaporation while the mixture is gently stirred under a stream of dry air or nitrogen until the mixture is completely dry. Fragile diatomaceous earth supports must be handled with extreme caution in this operation. Particles fragmented during stirring result in "fines" which decrease the permeability and performance of the column.

Superficially porous supports, being much more mechanically stable, can be handled with less caution. These materials may be coated by deposition of the stationary liquid from a volatile solvent by evaporation in a stream of dry nitrogen or air while slowly rotating in a round-bottom flask. The reproducibility by which liquid stationary phases may be deposited onto controlled surface porosity supports was discussed in a previous section.

Solvent Filtration Technique. Liquid chromatographic packings also can be prepared by the solvent filtration technique popular with some gas chromatographers. In this procedure, an excess amount of a 5-10% solution of the stationary liquid phase in a volatile solvent is poured onto the support contained in a flask. A vacuum then is placed on the solution several times to eliminate air from the support, degas the solution, and ensure complete wetting of the support by the solution. Excess solution is then filtered off and the support dried with gentle stirring while under a dry nitrogen or air stream or under gentle heat. Although this procedure produces satisfactory packings, the exact amount of stationary phase to be placed on the support cannot be accurately predetermined. In addition, unless this procedure is carried out in an exact manner, it is difficult to obtain packing with a reproducible amount of stationary liquid.

Coating of Prepacked Columns. Supports also can be coated with stationary liquid in prepacked columns (18). In this technique, the prepacked column containing support is filled completely with the stationary liquid dissolved in a volatile solvent. The solution is removed from the interstices of the packed bed by means of a slow downward stream of dry air or nitrogen. The remaining solvent is then evaporated by this gas stream. By using solutions containing different concentrations of partitioning liquid, one can vary the concentration of stationary phase that is loaded onto the support. The stationary liquid deposited in this manner is unevenly dispersed throughout the column. The column then is maintained in a constant-temperature environment (at elevated temperatures if required) for a period of time, during which homogeneous distribution of the stationary liquid phase takes place as a result of the gas-liquid equilibrium which is established.

Homogeneous dispersions of the stationary phase may
be obtained more rapidly if the column is evacuated to
about 1 mm to increase diffusivity with stationary
liquids having a vapor pressure of about 0.1 mm at room
temperature. This procedure is obviously limited to
liquids with a certain range of vapor pressure and,
although somewhat inconvenient, can produce columns with
excellent efficiency.

2. Column Loading

Dry-Packing Procedures. Narrow-bore chromatographic
columns can often be filled by dry-packing techniques,
providing the characteristics of the packing are suitable
(12). Dry packing may be carried out conveniently with
less dense, nonspherical particles, such as silica gel
and the diatomaceous earths, down to a particle size of
about 50 µ. Low-efficiency columns are generally obtained by dry-packing techniques if finer particles of
these materials are used. Spherical, dense, porous thin-
layer supports, such as Zipax® and Corasil®, may be homogeneously dry-packed down to a particle size of about
30 µ.

A useful technique for dry-packing straight columns
is not much different from that used to prepare high-
efficiency GC columns. A porous disk of sintered Teflon®,
nickel, or stainless steel (2-15 µ porosity, depending
on the particles) is forced into the internal diameter
of the outlet of a carefully cleaned column blank* to
retain the packing. Small amounts of the packing are
then introduced into the top of the vertically held column. The tubing is tapped vertically on the floor or
bench top, and the side of the tube is tapped (2-3 times
a second) at the level of the packing. A funnel may be
used to add the small incremental amounts of packing to

---

*The empty tubing may be cleaned by scrubbing the inter-
ior with Lakeseal® laboratory detergent solution, using
hot water and a long pipe cleaner, followed by washing
first with water and then with acetone, and drying with
a stream of nitrogen.

the column tubing. It is important that only about 0.5
cm of packing, usually 100-200 mg of material, be added
at a time, to ensure a homogeneous column. This addition
of small aliquots of packing is repeated until the entire
length of tubing is full. The column is then tapped for
an additional 5 or 10 min, to verify that it has been
completely filled.

The filled column is attached to the outlet of the
pump supplying the chromatographic carrier and pressurized, and air is eliminated from the packing. It is
desirable to increase the pressure to a value exceeding
that at which the column will subsequently be used and to
allow the carrier to flow for about half an hour. This
final treatment will verify whether or not the column was
packed to its maximum density. If not, the inlet of the
column will show a void due to settling of the packing.
If this is the case, the column has not been properly
prepared and should be remade.

After completing this packing process, a small plug
of quartz wool may be placed in the inlet of the column
to retain the packing. If in-stream rather than on-
column injection is to be employed, a plug of sintered
metal or porous Teflon® may be used in the inlet.

The technique described above is only one of several
that may be used to satisfactorily dry-pack a LC column.
Any technique which promotes the rapid establishment of
a dense, stable structure is suitable. For instance,
tamping of surface-etched glass beads with a glass rod
during the packing of columns has been recommended (19).

It should be noted that vibrating the chromatographic
column during the dry-packing procedure is usually undesirable. This is in contrast to the technique used by
many for the filling of GC columns. When filling a column of larger internal diameter by the technique described above, it is desirable to rotate the column
during the packing operation.

The time required for filling a 1-meter column, 2.1
mm i.d., is about 15-30 min, depending on the type of
support and its particle size. Smaller particles generally require longer packing times.

With proper care, highly reproducible columns of
Zipax® chromatographic support and other superficially
porous supports may be made by the technique described
above. Four replicate columns prepared from a master
batch of controlled surface porosity support packing
showed deviations in both retention times and plate
heights in the 2-5% (relative) range for two widely different carrier velocities (12). This dry-packing procedure also can be used satisfactorily for porous, nonspherical packings, such as the diatomaceous earths, as
long as the particle size is greater than about 50 μ.

Slurry-Packing Procedures. Some column packings, particularly those of very fine particle size, are put into
the columns with a slurry-packing procedure. In a wet-
packing process by sedimentation, segregation of particles
can occur because of differences in the settling velocity
of different sizes and densities, or as the result of the
generation of convection currents. This problem can be
overcome by using a filtration process (rather than sedimentation) for forming the packing. In a filtration
procedure, a suspension is forced to flow into a column
at a rapid rate by applying high pressure. As a result,
there is no chance for sedimentation or for unwanted convection currents counter to the flow of the particles,
as in sedimentation. A slurry-packing procedure is usually not practical with conventional LLC packings, however,
since the suspending liquids often have significant solubility for the stationary liquid phase.

A satisfactory procedure is to prepare a 10-25% slurry
of the chromatographic packing in a balanced-density solvent, such as 1,1,2,2-tetrabromoethane/Perclene® combination. This balanced slurry is then abruptly pumped
into the chromatographic column at a pressure which significantly exceeds that at which the column will be
subsequently used (20). The conditions for filling a
column with any specific packing material depend on the
density, particle size, and other properties of the
material itself. Columns with Permaphase® chromatographic packings (21) have been successfully prepared by this
approach. Unfortunately, conditions for the slurry

method of column packing have not been studied sufficiently to allow optimization of the technique. Results
are sometimes variable for reasons which have not been
established.

A different wet-filling approach sometimes used involves the consolidation of a loose-packing structure
with a binder, such as that used in thin-layer chromatography, the process of consolidation involving a wet-
filling technique with water as a suspending liquid (18).
This technique, which is also a filtration process, involves the filling of a stirred pressure vessel with a
mixture of packing material of active gypsum suspended
in distilled water at about 75°C. The suspension is
then pressured by nitrogen and kept homogeneous by rapid
stirring. Valves in the supply line at the lower end of
the column are opened, the suspension enters the column
(which has already been filled with distilled water),
and the bed starts to build up by filtration. When the
column is filled, the valves are closed and the column
is allowed to stand for about an hour to complete the
consolidation process. Stationary liquids can be introduced into the support by the coating procedure indicated for prepacked columns in the last section.

E. MOBILE-PHASE EFFECTS

1. Carrier Degassing

It is desirable to degas the carrier before it is
pumped through the chromatographic column. Air dissolved
in the carrier at high inlet pressures can subsequently
form small bubbles in the column eluate which seriously
disturb many detectors, particularly those using optical
properties. As a rough guide, the more polar the carrier, the greater is the tendency to dissolve air. A
convenient technique for eliminating the dissolved air
is to place a vacuum (about 20 in. of Hg, or better) on
a container of the carrier while stirring, until no evidence of degassing bubbles is seen. The container may
be warmed to hasten this degassing process. Under these
conditions, aqueous solutions, which have relatively high

solubility for oxygen, may be degassed in about 5 min.

Another reason for degassing the carrier is to remove
oxygen, which may react with the carrier or column stationary phase. An alternative method of removing dissolved oxygen is by vigorously sparging the carrier with
an inert gas such as nitrogen. Actually, the degassing
of oxygen is the main reason for the formation of the
gas bubbles in a detector. Sparging of the carrier is
desirable, even if the formation of gas bubbles in the
detector is not a problem.

An alternative approach for eliminating the formation
of small gas bubbles in UV or refractometer detectors is
to increase the pressure within the cell by 5-20 psi by
partially closing a valve located at the detector outlet.

2. Equilibration of Carrier with Stationary Phase

Since some mutual solubility of the two immiscible
liquids is always involved in LLC processes, special precautions must be taken to ensure long life of the chromatographic column. The carrier should be presaturated
with the stationary liquid to avoid gradual stripping of
this phase from the column during its use. Complete
equilibration of two immiscible liquids is actually more
difficult to achieve than might first be imagined. Since
there is limited solubility of the two phases, contact
between the phases in mixing is necessarily poor. A
technique which has been found to be satisfactory for
equilibrating the two phases is as follows. Place the
two phases (degassed or sparged with nitrogen) in a
closed container under nitrogen in the ratio of about
10 to 50/1 carrier to stationary phase. Rapidly mix
these components by magnetic stirring so that a turbid
vortex is formed, and continue the stirring for several
hours or overnight. The phases may then be separated;
the carrier phase is used for chromatography and the
stationary liquid recycled.

To ensure complete equilibration between the carrier
and the stationary phase in the chromatographic column,
it is desirable to use a precolumn arrangement. This
precolumn, attached in the flowing system just before

the inlet of the chromatographic column, provides for
complete equilibration of the carrier within the chromatographic column under the actual conditions at which
it is used. A satisfactory precolumn is a 50 cm x 6 mm
i.d. tube containing 20% of the stationary phase on a
finely divided (100-140 mesh) diatomaceous earth support,
such as that used in gas chromatography. This precolumn
may be changed frequently if necessary, without interfering with or altering the chromatographic column.

3. Temperature Effects

For optimum column stability, the carrier, precolumn,
and the chromatographic column should all be operated at
the same temperature. A stable system with variations
of less than 0.1°C is desirable. In addition to affecting column life, variations in temperature also can cause
significant changes in retention times, making qualitative identifications difficult and influencing the
precision of quantitative measurements.

4. Stability of Liquid-Liquid Chromatographic Columns

The useful life of a LLC column depends both on the
system and on the conditions under which the column is
operated. For maximum column life, one should pre-
saturate the carrier with stationary phase, eliminate
oxygen from the system, and use an adequate precolumn.
If these procedures are properly carried out, columns
often will last for months without significant change in
their chromatographic properties.

In some instances, the life of LLC columns is less
than expected. The reasons for this are several-fold
and will depend on the chromatographic system as well
as the equipment and technique used. Higher mutual solubility of the liquid phases generally means shorter
column life because of increased difficulty in maintaining proper equilibration.

The life of LLC columns, particularly those involving
superficially porous supports with very thin stationary
liquid films, can be greatly influenced by several
subtle factors. Chemical changes in the stationary

phase can require early replacement of the column. Such
an effect was noted in a system consisting of β,β'-oxy-
dipropionitrile as the stationary phase and dibutyl ether
as the carrier. Early studies with this system showed
variability in the solute retention times, sometimes
decreasing with column use, at other times increasing.
This variability was traced to two causes. First, it
was found that the carrier had not been fully saturated
with the stationary phase. As indicated previously,
less-than-saturated carrier can strip the stationary
phase from the support, resulting in shorter solute retention times as the column is used. Second, the increase in solute retention times was the result of
chemical changes within the column. Because of the
inadequacy of attempts to exclude air, the butyl ether
carrier was being oxidized during handling and storage
in the reservoir. Apparently peroxides formed in the
dibutyl ether react with β,β'-oxydipropionitrile, also
an ether, to produce a more polar stationary phase.
This increased polarity of the stationary phase results
in increased retention time of solutes. Rigorous exclusion of oxygen from the carrier system by vacuum degassing or by nitrogen purge effectively eliminated this
problem.

To minimize difficulties it is often desirable to re-
purify the carrier just before use. This is conveniently
accomplished by passing the solvent through a bed of
neutral or basic alumina (22). The reservoir containing
the carrier should also be continuously blanketed with
nitrogen to minimize oxygen pickup.

In spite of rigorous attempts to follow the steps previously described, one sometimes finds that systems with
certain LLC columns show continually decreasing solute
retentions. If pre-equilibration of carrier with stationary phase and no difficulties with an unwanted
chemical reaction (such as oxidation of the stationary
phase) are assumed, two additional reasons for the loss
of stationary phase during the use of LLC columns can
be postulated.

First, the solubility of the stationary phase in the
carrier might be influenced by pressure within the column. However, the very low order of compressibility of
liquids at the pressures at which liquid chromatography
is now conducted (up to about 5000 psi) probably pre-
cludes the significance of this effect.

The second effect which can remove stationary phase
from a LLC column in a system believed to be fully equilibrated involves an unsuspected temperature change within
the column packing. At nominal carrier velocities,
frictional forces can slightly increase the temperature
of the carrier within the chromatographic column (6).
These small, but significant, temperature increases
could result in a slow loss of stationary phase by carrier dissolution. Since the carrier velocity is smaller
in a precolumn of larger internal diameter than in the
chromatographic column, temperature increases within the
latter are larger than in the precolumn, resulting in a
loss of true equilibration between the carrier and the
stationary phase.

A simple experiment substantiated that there was indeed a significant temperature increase within a LLC
column operated at modest carrier velocities. A 1-meter,
2.1-mm-i.d. column containing a chemically bonded packing (6) was fitted with a microthermocouple in the
carrier stream at the outlet of the column, which was
at ambient temperature in air. After the initial temperature was observed, the carrier was passed through
the column with a pressure of 2400 psig, resulting in a
carrier velocity of 7.72 cm/sec and a flow rate of 5.55
cc/min. A temperature rise of about 1.6°C in the column
effluent was obtained under these conditions. From these
data one can extrapolate that a similar column operated
with a 500-psi pressure drop should involve a temperature rise of about 0.3°C, which is in agreement with a
theoretically predicted value. An increase of this
magnitude in the temperature of the packing can account
for a slow loss of stationary liquid during the use of
a LLC column. These results point out the desirability

of carefully thermostatting LC columns to avoid such
temperature changes.

At present it is not possible to state whether temperature or pressure effects are the primary cause of
the sometimes shorter-than-predicted life of LLC columns
in unusual cases. However, internal heating effects are
the more likely reason. If temperature effects are a
problem, a simple adjustment in chromatographic conditions eliminates the stationary-phase loss. The approach
is to slightly increase the temperature of the precolumn,
so that the carrier going into the chromatographic column is just supersaturated with the stationary phase.
Empirical adjustment of the level of supersaturation
hence allows the maintenance of the chromatographic column for months without change.

The stability of a LLC column can sometimes be affected adversely by recycling the carrier eluted from the
detector back into the reservoir for reuse. Generally,
recycling is not recommended, since changes can occur in
the carrier as the result of loss of equilibration or
the introduction of oxygen. (The amount of sample normally employed does not significantly degrade the UV-
transmitting properties of the solvent.)

F.   THE STATIONARY PHASE

1.   Liquid-Liquid Chromatographic Systems

At the present stage of development, the selection of
stationary phases for LLC is largely empirical. However,
the concepts presented in Chapter 4 provide a good basis
for choosing a particular system from an almost infinite
number of possibilities. In general, polar solutes require the use of polar stationary phases and relatively
nonpolar carriers. An example of this is shown in Figure 5.7. These polar substituted-aromatic urea herbicides are moderately retained by β,β'-oxydipropionitrile,
a highly polar stationary phase, using dibutyl ether as
the carrier.

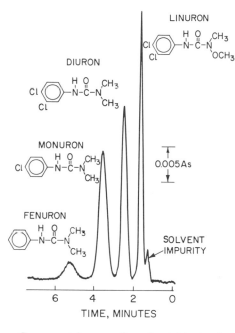

Figure 5.7. Separation of substituted urea herbicides.   Column = 50 cm, 2.1 mm i.d., 1.0% β,β'-
oxydipropionitrile on 37-44 μ controlled surface
porosity support; carrier = di-n-butyl ether;
flow rate = 1.14 cc/min; sample 1 μl of a solution of 67 μg/ml each in carrier. (Reproduced
from ref. 12 by permission of publisher.)

Nonpolar solutes may be separated most effectively by
using nonpolar liquid stationary phases with polar carriers. Figure 5.8 shows the separation of some relatively nonpolar fused-ring aromatics, using a hydrocarbon
polymer stationary phase with an aqueous methanolic
carrier. This form of "reversed-phase" liquid-liquid
chromatography often can be extended to somewhat more
polar solutes by increasing the concentration of water
in the carrier. An example of this effect is illustrated

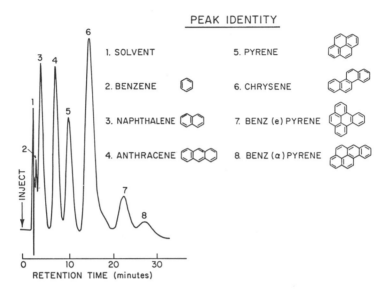

Figure 5.8. Separation of fused-ring aromatics by reversed-phase liquid chromatography. Column = 1 m, 2.1 mm i.d., hydrocarbon polymer on ≤37-μ Zipax®; carrier = 60% water/40% methanol (v/v); temperature = 40°C; column input pressure = 1200 psi; UV detector. (Reproduced from "Chromatographic Methods," 820M4, Mar. 30, 1970, by courtesy of John A. Schmit and by permission of the Instrument Products Division, E. I. du Pont de Nemours & Company, Wilmington, Del.)

in Figure 5.9, wherein a mixture of substituted anthraquinones was separated using the same hydrocarbon polymer stationary phase, but with a higher concentration of water. This form of chromatography is actually more generally applicable than was previously thought and should enjoy more widespread use in the future.

In theory, there are many combinations of binary immiscible phases with which LLC may be performed. However, practical limitations are involved in the selection of these liquids, as pointed out in Chapter 4. For instance,

PEAK IDENTITY

1. 9,10 ANTHRAQUINONE

2. 2-METHYL – 9,10 ANTHRAQUINONE

-3. 2-ETHYL – 9,10 ANTHRAQUINONE

4. 1,4 DIMETHYL – 9,10 ANTHRAQUINONE

5. 2-T-BUTYL – 9,10 ANTHRAQUINONE

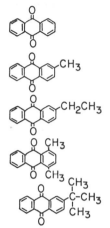

RETENTION TIME (minutes)

Figure 5.9. Reversed-phase separation of substituted anthraquinones. Column same as in Figure 5.8;
carrier = 50% water/50% methanol (v/v); column
pressure = 900 psi; UV detector. (Reproduced from
"Chromatographic Methods," 820M4, Mar. 30, 1970,
by courtesy of John A. Schmit and by permission of
the Instrument Products Division, E. I. du Pont de
Nemours & Company, Wilmington, Del.)

if an ultraviolet detector is employed, UV-transmitting
carriers and stationary phases also must be used.

The number of stationary phases needed for LC separations need not be as great as might be anticipated.
Since the partition ratio of solutes can be drastically
affected by the polarity of the carrier, large changes
in relative retentions may be gained by minor changes in
the nature of the carrier phase. For instance, if hexane
is used as a carrier and the solutes to be separated are
strongly retained on the column, the desired retention
times often may be obtained by adding small amounts of
a polar modifier, such as chloroform or tetrahydrofuran,
to the carrier. By establishing the proper concentration of polar modifier in the nonpolar carrier, one may
obtain optimum partition ratios for the separation (see
Chapter 4 for a more detailed discussion).

Some examples of LLC systems are given in Table 5.3.

Table 5.3

Illustrative Liquid-Liquid Chromatographic Systems

Stationary Phase	Carriers
β,β'-Oxydipropionitrile	Cyclopentane, hexane, hep-tane, isooctane
Carbowax® 600	Same, modified with up to 10% chloroform, dichloro-
Triethylene glycol	methane, tetrahydrofuran, dioxane
Ethylene glycol	Di-n-butyl ether
Dimethyl sulfoxide	Isooctane
H₂O/ethylene glycol	Hexane/CCl₄
Ethylenediamine	Hexane
Hydrocarbon polymer	Aqueous methanol
Chloroform/cyclohexane/ nitromethane	
Ternary mixture, two phases	

This list is by no means exhaustive, but the systems described are indicative of the types which may be used for desired separations. Highly polar ethers or hydroxyl-containing liquids, such as β,β'-oxydipropionitrile and triethylene glycol, may be paired with a variety of non-polar carriers. For less polar solutes, the most non-polar carriers, such as hexane and cyclopentane, should be employed. When working with solutes of higher polarity, one may modify these nonpolar carriers with polar solvents, like chloroform and tetrahydrofuran, up to a concentration of about 10% by volume. Alternatively, a more polar single carrier (e.g., di-$\underline{n}$-butyl ether) may be employed. Stationary phases with other highly polar functional groups, such as dimethyl sulfoxide, often can be used advantageously with carriers like hexane and isooctane. Hydroxylic stationary phases containing water may be combined with nonpolar or immiscible partially polar carriers.

Ternary or quaternary systems may be mixed to form two immiscible phases useful for liquid-liquid partition. An example is the chloroform/cyclohexane/nitromethane system. Certain amines, such as ethylenediamine, are useful in conjunction with nonpolar carriers. As illustrated previously, nonpolar hydrocarbon polymers may be employed with aqueous alcoholic carriers for "reversed-phase" chromatographic separations. A series of liquid-liquid partition systems which may find use in LLC has been described by Metzsch (23).

The use of monomers as stationary phases is preferred over polymeric materials for several reasons. Because of differences in the properties of polymer samples due to molecular weight variations, one might also expect a resulting variability in the chromatographic properties. Also, the variability of polymers makes difficult the maintenance of a carrier which is fully saturated with a constant species from the stationary phase. In addition, the differing stabilities of variable polymeric materials to chemical attack, oxidation, etc., could lead to unwanted variations in retention times. Finally, columns prepared with stationary phases of higher

molecular weight might be expected to have lower efficiencies, as discussed below.

The viscosity of the stationary liquid can have a
significant effect on column efficiency. Since columns
with superficially porous supports are limited by mass
transfer in the mobile phase but generally not in the
stationary phase, the use of viscous liquid phases with
small diffusion coefficients would not normally be expected to affect efficiency. However, stationary phases
with higher viscosity provide more opportunity for
coated support particles to adhere to one another at
contact points, resulting in a loss in column packing
regularity and lower column efficiency (19). Therefore,
the use of viscous (polymeric) stationary phases, such
as those widely employed in GLC, should be avoided if at
all possible.

2. Surface-Reacted (Bonded) Stationary Phases

As previously indicated, conventional liquid-liquid
chromatography has a significant procedural limitation.
Since there is usually some solubility of the stationary
phase in the mobile liquid, the latter must be presaturated with the stationary liquid to avoid gradual stripping
of this phase from the column. In addition, the rather
high flow rates used in high-speed separations sometimes
generate in narrow-bore columns shear forces which tend
to remove the stationary liquid from the support.

To overcome these disadvantages, column packings with
chemically bonded stationary phases have been developed.
Experimentally, these packings have distinct advantages,
since no presaturation of the carrier and no precolumns
are required and no pretreatment or modification of the
packing is necessary. These new chemically bonded stationary phases are currently available in two different
forms, as described below.

Esterified Siliceous Supports. The esterification of
siliceous supports with a monomolecular organic layer of
alcohols produces chromatographic packings with a chemically bonded stationary phase. The esterification of
porous glass, Porasil® C, with alcohols has resulted in

phases called "brushes", with useful gas and liquid chromatographic properties (24). These materials are now
commercially available as Durapak® from Waters Associates.
An example of a LC separation made with a Durapak® column is shown in Figure 5.10. This column packing shows
excellent resolution for the components in the sample;
however, column efficiency is relatively low at the
carrier velocity used, presumably because of the completely porous nature of the packing (see Chapter 9 for
a separation with "brushes" on a porous thin-layer
support).

Separations afforded by esterified siliceous supports
probably do not involve liquid-liquid interactions, but
are really adsorption chromatography at the monomolecular
layer of organic modifier, in conjunction with the residual activity of any remaining surface silanol groups.
Unfortunately, silicate esters have poor hydrolytic and
thermal stability, which limits the utility of these
materials.

Supports with Chemically Bonded Silicone Polymers.
Column packings with chemically bonded, nonextractable,
thermally and hydrolytically stable, polymeric silicone
stationary phases have been prepared which show important
LC (and GC) characteristics (21,25). One form of these
bonded-phase packings is now available from E. I. du
Pont de Nemours & Co. as Permaphase® chromatographic
packings. These materials are prepared by reacting
silane reagents with the surface of the porous shell of
Zipax® support and then polymerizing the reagents to give
the desired silicone coating (21). The polymeric coat-
ings may be prepared with a variety of functional groups,
ranging from very polar to nonpolar, resulting in widely
diverse selectivities. The physical characteristics of
the polymeric stationary phase, such as concentration,
film thickness, and structure (linear, cross-linked,
etc.), may be controlled to obtain the desired chromatographic properties. The unique surface of these bonded-
phase materials, particularly those containing the
highly polar functional groups, provides unusual selectivities.

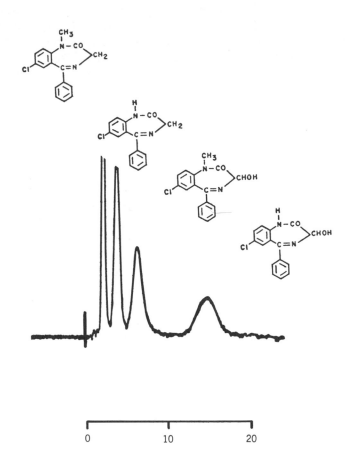

Figure 5.10. Separation of some benzodiazepines with Durapak®-OPN. Column = 1 meter, 1 mm i.d., Durapak®-OPN, 36-75 μ; carrier = hexane/isopropanol (80-20 v/v); flow rate = 1.0 cc/min; room temperature: sample = approx. 8 μg total; LDC model 1205 UV Monitor. (Reproduced from C. G. Scott and P. Bommer, J. Chromatog. Sci. 8, 446 (1970) by permission of authors and publisher.)

Chromatographic packings with chemically bonded silicone phases provide excellent column efficiency and stability, and eliminate the problems associated with

the loss of partitioning phase from the support during
the operation of conventional LLC columns. In addition,
many highly polar carriers, such as chloroform and
alcohol-water mixtures, can be used to elute strongly
retained sample components rapidly without effecting a
change in the properties of the chromatographic column.
Figure 5.11 shows the separation of highly polar substituted acetanilides and sulfonamides, carried out on a

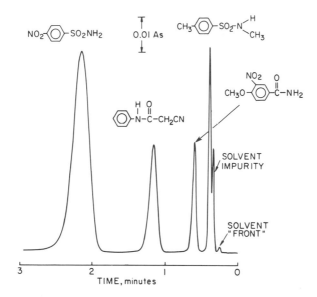

Figure 5.11. Separation with chemically bonded
"nitrile" phase column. Column = 1 m, 2.1 mm
i.d., 0.75% bonded-"nitrile" polymer on <37-μ
Zipax® support; carrier = 1:1, 2-chloropropane/
hexane; flow rate = 4.35 cc/min; column input pressure = 1360 psi; temperature = 27°C. (Reproduced
from ref. 21 by permission of publisher.)

chemically bonded "nitrile" column using a carrier of
1:1 2-chloropropane/hexane. Use of this very polar
carrier in conventional LLC is questionable because of
its high solubility for most liquid stationary phases.
It appears that packings with chemically bonded organic

phases will assume major importance in LC for a large variety of separations.

G. APPLICATION CONSIDERATIONS

1. General Experimental Approach for Analysis

It is desirable to briefly review the procedure by which the parameters for a particular LLC separation are selected.

a. The chromatographic system should be selected so that the solute has relatively high retention in the stationary liquid phase and relatively low solubility in the carrier. Figure 5.12 shows the separation of some phenolic compounds, using an "ether" bonded-phase

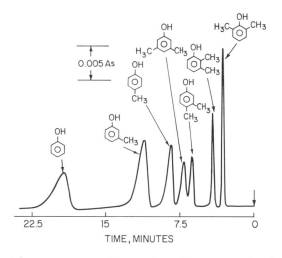

TIME, MINUTES

Figure 5.12. Separation of substituted phenols. Column = 1 m, 2.1 mm i.d., 0.88% "ether" bonded-phase < 37 μ; carrier = 2.5% methanol in cyclo-pentane; flow rate = 1.00 cc/min; inlet pressure = 250 psi; temperature = 27°C; sample = 20 μl of 0.25 mg/ml p- and m-cresol, 0.125 mg/ml each of phenol and 2,6-, 2,3-, 3,4-, and 3,5-dimethyl-phenol. (Reproduced from J. Chromatog. Sci. by permission of publisher.)

column. This packing has strong electron-donating

tendencies which selectively retard compounds that
hydrogen-bond.

  b. The parameters should be adjusted so that solute
peaks elute in the optimum partition ratio range, which
is generally 2-5, as indicated in previous chapters. In
this partition ratio range, optimum resolution of peaks
(as a function of the time of analysis) is obtained. In
addition, peaks are relatively sharp, providing good
sensitivity of detection. The optimum partition ratio
range is obtained conveniently by adjusting the carrier
polarity. An example of the importance of this parameter
is shown in Figure 5.13, which represents the separation

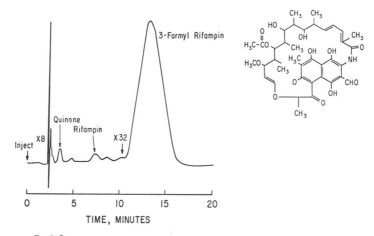

Figure 5.13. Separation of impurities in 3-formyl-
rifampin. Column = 1 m, 2.1 mm i.d., polyamide
on < 37-μ Zipax®; carrier = 75% hexane/25% ethanol
(v/v); UV detector. (Reproduced by courtesy of
John A. Schmit, unpublished studies.)

of some impurities in a 3-formylrifampin sample. It was
necessary to determine that the optimum carrier for this
separation with a Zipax®/polyamide column was 75% hex-
ane/25% ethanol. This combination permits low-level
detection of the desired impurities and at the same time

provides good measurement of the very polar, high-
molecular-weight major constituent.

c. Where possible, carriers and stationary phase per-
mitting the highest diffusion coefficients should be
employed to optimize column efficiency.

d. The resolution of the sample components should be
determined primarily by careful selection of a liquid-
liquid system having the greatest separation factors for
the solutes in the mixture; however, column efficiency
should be optimized so that the analysis can be carried
out as rapidly as possible with a high degree of peak
sharpness for maximum detection sensitivity.

2. Applications of Liquid-Liquid Chromatography

Quantitative Analysis. Precise quantitative analysis
can be routinely carried out by LLC (26). Quantitation
is obtained by adding an internal standard to the sample
and carrying out area ratio measurements, using techniques which are common to gas chromatography. When an
electronic integrator is employed to measure the peak
areas of separated components, LC analyses with a standard deviation of less than 0.5%, relative, are not
uncommon (26). (The use of high-pressure sampling
valves, which ensure the reproducible introduction of
very small sample aliquots, may eliminate the need for
the internal standards used in quantitative analysis.)
These results do not represent the limit of precision at
which analyses may be carried out by LLC. Analyses with
reproducibilities equal or superior to the best reported
for GC might be expected with the use of sophisticated
instrumentation and computer read-out.

Trace Analysis. Analysis of components present in
trace quantities can also be accomplished by LLC, although the sensitivity of the measurement is highly
dependent on the detector used. The ability of LLC columns to handle relatively large sample volumes without
significant peak broadening often compensates for a limited detector response. The ability to use a large volume of sample, for example, 100 μl with a 1-meter, 2.1-
mm-i.d. column, permits the measurement of low-level

components. Some UV photometric detectors have sufficient sensitivity to permit the measurement of peaks in
the nanogram or picogram range. Figure 5.14 shows that
10 ng of diuron 3-(3,4-dichlorophenyl)-1,1-dimethyl-
urea, absorptivity = 88 liters/g-cm at 254 nm is easily

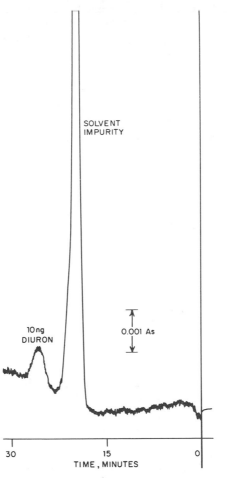

Figure 5.14. High-sensitivity UV detection. Carrier = di-n-butyl ether; flow rate = 0.26 cc/min.
(Reproduced from ref. 27 by permission of publisher.)

detected when a 100-μl sample aliquot is injected into
the chromatographic column (27). High-speed LLC has been
used for the analysis of trace residues of an organophos-
phorous larvicide in surface waters (28).

Flow Programming. Carrier flow programming can sometimes be advantageously employed in LLC separations.
Flow programming in conjunction with detectors working
on a bulk property principle is difficult; however, this
technique, used with solute-property detectors, such as
UV photometry and flame ionization, can reduce significantly the time required for an analysis. Superficially
porous or porous thin-layer supports are preferred in
flow programming studies, since columns of these materials exhibit the least decrease in column efficiency as
carrier velocities increase.

Temperature Programming. This technique is generally
not recommended for conventional LLC. There is considerable experimental difficulty in maintaining equilibration
between the carrier and the stationary phase, so that the
stationary liquid on the column will not "bleed." The
benefits which may be expected from temperature programming are marginal and usually do not justify the experimental difficulties involved.

Gradient Elution. This versatile technique is not
practiced in conventional LLC. The increased polarity
during the gradient run means greater solubility of the
carrier for the stationary phase, resulting in increasing
column "bleed" as the polarity of the carrier rises. However, the development of packings with chemically bonded
stationary phases has made this technique feasible. Since
the bonded stationary phases are not removed by the change
in the polarity of the carrier, these packings can be used
in systems in which wide variations in the polarity of the
carrier are employed. A particular advantage of the
bonded-phase packings for gradient elution is that no pre-
equilibration of the carrier with the stationary phase
(or with a modifier such as water, in the case of adsorption chromatography) is required, which constitutes a
distinct improvement in the convenience of carrying out
routine analyses.

Very-High-Speed Analyses. Analyses may be carried out
in very short times if columns provide large separation
factors while operating at high efficiency. Columns with
superficially porous or porous thin-layer packings generally offer the greatest potential in this area, as discussed in Chapter 1. Figure 5.15 shows the separation
of methoxychlor from an analog in less than 60 sec with
a resolution of 2.8. The same column, 24 cm in length,

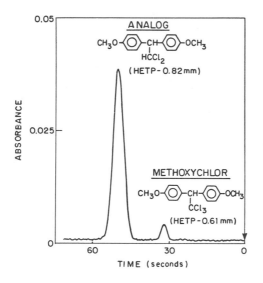

Figure 5.15. High-speed liquid-liquid chromatographic separation. Column = 50 cm, 3.2 mm i.d.,
0.5% β,β'-oxydipropionitrile on 37-44 μ controlled
surface porosity support; carrier = hexane; carrier flow rate = 3.2 cc/min; temperature = 27°C;
column inlet pressure = 115 psi; sample = 4 μg
total. (Reproduced from ref. 7 by permission of
publisher.)

would separate these compounds to base line (R_s = 1.5)
in 8 sec.
Process (On-Stream) Analysis. The rapid repetitive
analyses needed for on-stream process analysis by LLC now
appears to be practical. A prime prerequisite for this
type of work - the availability of very stable, high-

performance columns - has been met with the development
of chemically bonded packings. It is anticipated that
developments in LLC process analysis will rapidly take
place.

    <u>Preparative Liquid-Liquid Chromatography</u>. Larger-
scale separations by LLC are feasible, but more information about the performance of columns with large internal
diameter is needed before approaches in this area can be
optimized. Milligram quantities of components may be
isolated from larger-diameter columns at very high column
efficiencies, using the superficially porous packings designed for high-speed chromatography, even though these
materials have limited sample capacity (<u>17</u>). For increased capacity, columns containing packings with high
surface area should be employed in conjunction with relatively slow carrier velocities.

    With the conventional LLC approach, the isolated frac-
tion contains a saturation concentration of the stationary phase as a contaminant. This contaminant has to be
removed in some manner before a pure fraction can be obtained. With chemically bonded silicone stationary
phases, contamination from packings usually is absent.

References

1. A. J. P. Martin and R. L. M. Synge, Biochem. J.,
    <u>35</u>, 91 (1941).
2. J. C. Giddings and R. E. Keller, in Vol. 1 of "Chro-
    matographic Science Series"; "Dynamics of Chromatog-
    raphy," Part I, "Principles and Theory," J. C.
    Giddings and R. E. Keller, eds., Marcel Dekker, New
    York, 1965.
3. J. C. Giddings, "Theoretical Basis of Partition Chro-
    matography," in "Chromatography," 2nd ed., E.
    Heftman, ed., Reinhold, New York, 1967.
4. J. F. K. Huber, "Liquid Chromatography and Columns,"
    in Vol. 2B, "Comprehensive Analytical Chemistry,"
    C. L. Wilson and D. W. Wilson, eds., Elsevier,
    Amsterdam, 1968.

5. D. C. Locke, "Thermodynamics of Liquid-Liquid Par-
    tition Chromatography," in Vol. 8 of "Advances in
    Chromatography," J. C. Giddings and R. A. Keller,
    eds., Marcel Dekker, New York, 1969.
6. J. J. Kirkland, "Some Current Aspects of High-Speed
    Liquid Chromatography," The Gordon Research Confer-
    ence in Analytical Chemistry, New Hampton, N. H.,
    August 12, 1969.
7. J. J. Kirkland, Anal. Chem., 41, 218 (1969).
8. J. J. Kirkland, U.S. Patent 3,505,785.
9. L. R. Snyder, J. Chromatog. Sci., 7, 352 (1969).
10. F. A. van Niekerk and V. Pretorious, in "Advances
    in Gas Chromatography, 1967," A. Zlatkis, ed., Pres-
    ton Technical Abstracts Co., Evanston, Ill., 1967,
    p. 167.
11. J. J. Kirkland, unpublished studies.
12. J. J. Kirkland, J. Chromatog. Sci., 7, 7 (1969).
13. R. E. Majors, J. Chromatog. Sci., 8, 338 (1970).
14. B. L. Karger, H. Engelhardt, and K. Conroe, Eighth
    International Symposium on Gas Chromatography,
    Dublin, Sept. 28-Oct. 1, 1970.
15. J. H. Knox and J. F. Parcher, Anal. Chem., 41, 1599
    (1969).
16. S. T. Sie and N. van den Hoed, in "Advances in Chro-
    matography, 1969," A. Zlatkis, ed., Preston Techni-
    cal Abstracts Co., Evanston, Ill., 1969, p. 318.
17. J. J. DeStefano and H. C. Beachell, J. Chromatog.
    Sci., 8, 434 (1970).
18. S. T. Sie and N. Van den Hoed, J. Chromatog. Sci.,
    7, 257 (1969).
19. B. L. Karger, K. Conroe, and H. Engelhardt, J.
    Chromatog. Sci., 8, 242 (1970).
20. J. J. Kirkland, paper submitted to J. Chromatog.
    Sci.
21. J. J. Kirkland and J. J. DeStefano, J. Chromatog.
    Sci., 8, 309 (1970).
22. G. Hesse, I. Daniel, and G. Wohlleben, Angew. Chem.,
    64, 103 (1952).
23. F. A. V. Metzsch, Angew. Chem., 65, 586 (1953).

24. I. Halász and I. Sebestian, Angew. Chem., Intern.
     Ed., <u>8</u>, 453 (1969).
25. W. A. Aue and C. R. Hastings, J. Chromatog., <u>42</u>,
     319 (1969).
26. J. J. Kirkland, J. Chromatog. Sci., <u>7</u>, 361 (1969).
27. J. J. Kirkland, Anal. Chem., <u>40</u>, 391 (1968).
28. R. A. Henry, J. A. Schmit, and J. F. Dieckman, Sym-
     posium on Liquid Chromatography, American Chemical
     Society Fall Meeting, Chicago, Ill., Sept. 15, 1970.

CHAPTER 6

The Practice of Liquid-Solid Chromatography

Lloyd R. Snyder

A. INTRODUCTION

"Modern" liquid chromatography includes four distinct families of procedures: liquid-solid (adsorption), liquid-liquid (partition), ion-exchange, and exclusion (gel permeation and gel filtration). Liquid-solid chromatography is described in this chapter, with the remaining methods being covered in three other chapters.

Of these four LC procedures, liquid-solid chromatography is potentially one of the most useful. Until recently, however, modern LSC has found quite limited practical application in comparison to the other three methods. To a large extent this has been due to the great success of thin-layer chromatography (TLC), which can be used for the same types of samples that can be separated by LSC in columns (i.e., modern LSC). Thin-layer chromatography can be simple and inexpensive, and it often provides completely adequate separation of fairly complex mixtures. At the present time there also exist a large number of workers with the necessary experience to obtain consistently good results by TLC.

Liquid-solid chromatography, on the other hand, has a number of potential advantages relative to TLC:

- Greater speed and separation efficiency (see, e.g., ref. 1).
- Easier automation for convenience and unattended operation.
- Easier quantitation.
- Easier scale-up for preparative separations.
- Applicability to process control.

As more workers become capable of exploiting these advantages (i.e., gain the necessary experience), LSC will become as widely used as TLC.

How is a particular LSC procedure chosen from among the four possible LC methods? Specifically, in what

205

situations should the first choice be LSC?   Several factors which relate to this question come to mind:

- Experimental convenience.
- Personal experience with a particular LC method.
- Sample type.
- Unique selectivity.

Convenience and relative experience with a given method
are not compelling reasons for the selection of one procedure over another. Sample type is usually the dominant
consideration. Liquid-solid chromatography is well suited
to many samples of intermediate molecular weight (i.e.,
less volatile samples with molecular weights <1000), particularly those which are oil-soluble. Any sample type
which has been successfully separated by TLC (e.g., see
ref. 2) can be separated by LSC. This includes most types
of compounds, from nonpolar hydrocarbons to fairly polar,
water-soluble compounds. For high-molecular-weight (e.g.,
polymeric) samples or ionic, lipophobic compounds, LSC is
less promising.

Difficult separations may require a specified kind of
selectivity, that is, $\alpha = 1$; see Chapter 1. Liquid-solid
chromatography tends to exhibit a higher selectivity among
certain isomer types than do other LC methods. It also
provides sharp discrimination among different types of
compounds, that is, compounds differing in the kind or
number of functional group substituents. On the other
hand, LSC tends to ignore differences in alkyl substitution or compound molecular weight. The next section provides a more detailed description of selectivity in LSC,
along with a discussion of the origins of this selectivity.

B. SELECTIVITY IN ADSORPTION CHROMATOGRAPHY

Separations by LSC are usually carried out on polar
adsorbents, such as silica, alumina, or other inorganic
solids. The primary factor in determining the relative
adsorption of a sample molecule (i.e., its k' value) is
its functional groups. Relative adsorption increases as
the polarity and number of these functional groups

increase, because the total interaction between the molecule and the polar adsorbent surface is thereby increased. However, the same can also be said for liquid-liquid (partition) chromatography, where an increase in the number and polarity of the groups in a sample molecule increasingly favors its retention in the polar liquid phase (usually the stationary phase). The uniqueness of retention and selectivity in LSC arises from two characteristic features of adsorption from solution: (1) a competition between sample and solvent molecules for a place on the adsorbent surface, and (2) multiple interactions between functional groups on the sample molecule and corresponding rigidly fixed sites on the adsorbent surface.

The competitive nature of the adsorption process in LSC was referred to in Chapter 4. The adsorption of a sample molecule, X, from the mobile phase, m, proceeds by the displacement of some number n of solvent molecules, M, that are initially in the adsorbed phase, a:

$$X_m + nM_a \rightleftharpoons X_a + nM_m$$

That is, the surface of a solid (the adsorbent) in contact with a liquid (the solvent) is completely covered by solvent or sample molecules. Adsorption of X requires the desorption of sufficient solvent molecules to permit the accommodation of X on the adsorbent surface. The net energy of adsorption of X, which determines its relative adsorption and k', is therefore determined by the <u>difference</u> in adsorption energies of individual functional groups and corresponding solvent molecules which are displaced by these groups upon adsorption. In the case of nonpolar (weak) solvents, the energy of adsorption of an alkyl group substituted onto X will be approximately canceled by the adsorption energy of the solvent. With stronger solvents the net energy of adsorption of an alkyl group will be negative, that is, unfavorable for adsorption. In the latter case the alkyl group then dangles out from the adsorbent surface (i.e., partially desorbs), even though the remainder of the sample molecule, X, is adsorbed. In either case--weak or strong solvents--the net result is little or no contribution of an alkyl group

to the net adsorption energy (and k') of X. This means
that LSC normally shows little selectivity among homologs,
that is, no molecular weight selectivity. Rather, there
is a pronounced tendency toward compound type selectivity,
which leads in turn to the possibility of compound type
separations, as in the analysis of lipids (3) and petroleum (4).

Table 6.1 illustrates the kind of compound type breakdown which is thus possible. Here all compounds possess-
ing a particular functional group or a certain number of

Table 6.1

Compound Type Separations by LSC of
Lipids or Petroleum

A. Lipids (3)

    Hydrocarbons
    Cholesterol esters $(R-CO_2-)$
    Triglycerides   (three $R-CO_2-$ groups)
    Free sterols   (-OH)
    Diglycerides   (two $R-CO_2-$ groups plus -OH)
    Monoglycerides   ($R-CO_2-$ plus two -OH groups)
    Free fatty acids   (-COOH)

B. Petroleum (4)

    Saturated hydrocarbons
    Monoaromatics (alkyl benzenes, thiophenes)
    Diaromatics (alkyl naphthalenes, benzothiophenes)
    Triaromatics (alkyl phenanthrenes, anthracenes, etc.)
    Tetraaromatics
    Polar heterocompounds (phenols, carbazoles, etc.)

---

the same functional group are lumped into a single fraction, regardless of differences in alkyl or cycloalkyl
substitution.

The rigid adsorbent fixes the positions of the reactive groups or adsorption sites on its surface. The interaction of corresponding functional groups in a sample
molecule with these sites will vary with the geometry of

the molecule, being stronger when the positions of groups
and sites are appropriately matched. As a result the
relative adsorptions of different isomers often vary markedly, in relation to corresponding differences in retention in liquid-liquid systems. In the liquid (solution)
phase solvent molecules are free to align themselves with
sample molecules for maximum interaction between their respective groups. Consequently, as already mentioned, LSC
usually provides more pronounced selectivity among isomers than is found for the other LC methods.

Table 6.2 provides some examples of the types of separations which are readily accomplished by LSC. Often small
differences in structure lead to large differences in α.
For example, on alumina the k' value of compound I can be
made 20 times larger than that of compound II. It would
be difficult to imagine similar α values for any other LC
method.

Special selectivity in LSC can frequently be achieved
by varying the solvent composition; Table 4.4 of Chapter
4 provides one example, and a number of others are sum-
marized in ref. 5. Although the same can be said of ion-
exchange and liquid-liquid chromatography, the number of
possible solvent systems (which are chromatographically
useful) is much larger in LSC, and it is easier to explore the selectivity of different solvent systems with
this technique. The potential selectivity of different
solvent systems in LSC also tends to be greater, because
of the preferential concentration of polar solvent components (from solvent mixtures) into the adsorbed monolayer.
This leads to an enormous variety of possible stationary
phases (i.e., adsorbed solvent plus adsorbent) and a corresponding range in potential selectivities.

For a detailed discussion of selectivity in LSC, see
ref. 5.

Table 6.2

Isomer Separations by LSC (Alumina as Adsorbent) (6-8)[a]

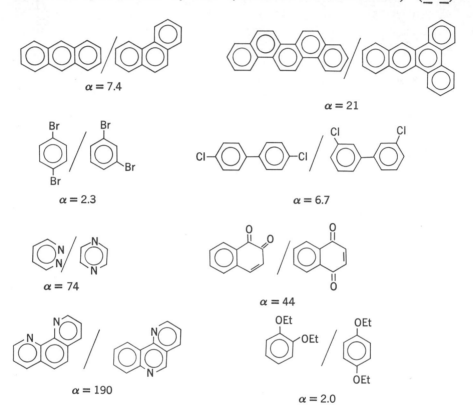

[a] In some cases α varies markedly with solvent composition; reported α values are usually the largest that have been obtained.

C.   PRACTICAL CONSIDERATIONS IN LIQUID-SOLID
         CHROMATOGRAPHIC SEPARATIONS

1.   Selecting a Column

The first problem in designing an adequate LSC separation is the selection of a column. We must specify the type of adsorbent (silica, alumina, etc.); its particle size, surface area, and geometry (spherical versus irregular, porous versus porous thin-layer or pellicular, etc.); the water content of the adsorbent; the size of the column; and the column-packing procedure. The following discussion of these various factors is necessarily brief. For a more complete treatment, see refs. 1 and 5.

Usually we do not have to worry about adsorbent type. Silica is a good, general-purpose adsorbent which is commercially available in a wide variety of useful forms. In addition, silica is nearly optimum in several respects: little or no reactivity toward most sample types (see later discussion), high linear capacity (i.e., constant sample retention volumes for large sample charges), and high efficiency. Only when a change in separation selectivity is required, and when changes in the solvent are unable to provide this change in selectivity, should the possibility of another adsorbent type be considered. This might be the case for unusually difficult separations, whether of very similar compounds or of mixtures containing a large number of components. For example, the separation of polycyclic aromatic hydrocarbons is better carried out on alumina than on silica, because the α values for adjacent benzologs (e.g., naphthalene and phenanthrene) are much greater on alumina than on silica, and isomers of these compounds (e.g., anthracene and phenanthrene) are more readily separated on alumina (see Table 6.2). Similarly, unique selectivity is known to be associated with many adsorbents (see the discussion in ref. 5), and in some cases the general advantages of silica can be overshadowed by the greater selectivity of another adsorbent for a particular separation.

The sequential use of more than one adsorbent type (or more than one LC method) should be considered for very complex mixtures, such as certain natural products and petroleum. An example is provided in Figure 6.1 for the separation of petroleum heterocompounds (compounds containing nitrogen or oxygen). Here three different adsorbents (silica, alumina, and charcoal) are combined with ion-exchange chromatography to effect the sequential isolation of 36 different compound types. The initial petroleum distillate is first separated on alumina to remove

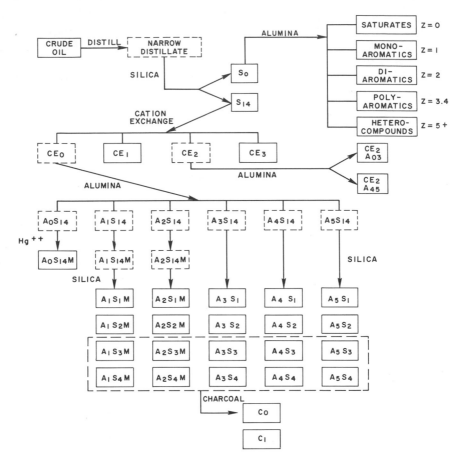

Figure 6.1. Scheme for the separation of nitrogen-
or oxygen-containing compounds in petroleum (9).
(Reprinted from Accounts Chem. Res. by permission of
the American Chemical Society.)

hydrocarbons and most sulfur compounds from the nitrogen
and/or oxygen heterocompounds. The heterocompound concentrate S_{14} is then further separated into basic, CE_1-
CE_3, and nonbasic, CE_0, fractions by cation-exchange
chromatography. The nonbasic fraction, CE_0, is next separated sequentially on alumina, Hg^{+2}-impregnated cation-
exchange resin, and silica to yield several different

fractions differing in polarity and acidity. Finally,
certain of the latter fractions are separated on charcoal
to yield aliphatic, C_0, and aromatic, C_1, heterocompound
types.

The surface area of an adsorbent has an important effect on its chromatographic properties. To a first approximation, sample retention volumes and adsorbent linear capacity are each proportional to the specific surface
area of the adsorbent. Commercially available porous silicas have surface areas between 300 and 800 m^2/g; porous
aluminas, between 100 and 200 m^2/g. Pellicular adsorbents have much smaller surface areas; examples are Cora-
sil®-I and Corasil®-II with 7 and 14 m^2/g, respectively.
These latter adsorbents have undesirably low linear capa-
cities, which preclude their use in preparative work and
lead to an effective loss in detection sensitivity. However, these disadvantages are offset by higher efficiencies for the pellicular adsorbents (see below).

The size and geometry of the adsorbent particle play
a critical role in separation, as discussed in Chapter 1
in general terms. A detailed discussion for LSC has been
reported recently (1). We have seen that "good" columns
have large numbers of theoretical plates N, at relatively
high solvent flow rates, F, without excessive pressure
drops, ΔP, across the column. Since N, F, and ΔP are
interrelated in a somewhat complex fashion, it is useful
to compare different columns in terms of total number of
plates for a given time and pressure drop. We will follow this procedure in the present discussion (see Table
6.3).

The effect of particle size and geometry on column
efficiency, N, is related to how the column is packed.
Irregular, porous silica (i.e., ordinary silica gel) can
be satisfactorily dry-packed if the average particle diameter is ≥ 50 μ. An optimum column (largest N value) is
obtained simply by adding small portions of silica (no
solvent) to the column, and tapping the sides of the column for maximum settling between each addition. As seen
in Table 6.3 (50-μ and 100-μ particles), 50-μ particles
give significantly better columns than larger particles,

for a given set of experimental conditions (time, pressure drop, k'). The higher permeability of 100-μ particles cannot overcome their larger H values. Furthermore, this is achieved with shorter, more convenient columns (8 meters versus 20 meters in Table 6.3). Slightly better columns can be obtained with spherical porous silica (Waters' Porasil® A in Table 6.3) of similar particle size (35-75 μ). In the author's experience this material is the best general-purpose silica available for LSC. Smaller silica particles (<40 μ), either irregular or spherical, give poor columns by the dry-packing procedure. However, these small-diameter silicas yield very good columns by the pressure slurry-packing procedure (1); see the 20-μ column of Table 6.3. Unfortunately, such columns are at present difficult to prepare; silicas of small particle diameter are not commercially available in narrow mesh ranges, and the pressure slurry-packing technique has not been studied sufficiently to guarantee consistently good columns. Performance equivalent to that of these small-particle columns can be obtained with pellicular silicas such as Corasil®-II (see Table 6.3).

Table 6.3

Comparative Performance of Different Silica Columns Based on 1000-psi Pressure Drop, 10-min Separation Time, k' = 3

Column	N, plates[a]	L, meters
100-μ irregular silica, dry-packed	750	20
50-μ irregular silica, dry-packed	900	8
Porasil® A, 35-75 μ	1000	3.5
20-μ irregular silica, pressure slurry-packed	1200	1
Corasil®-II, dry-packed	1200	4

aThe plate numbers are approximate and are intended mainly to show trends and rough values for typical columns.

The latter adsorbent provides about the same column efficiency as a 20-μ pressure slurry-packed column. The
pellicular adsorbents such as Corasil®-II (with particle
diameters of about 50 μ) are easily dry-packed as described above, but the sample sizes that can be tolerated
are only one-tenth to one-twentieth as great as those for
porous silicas (see above). Another disadvantage of the
pellicular adsorbents is their expense.

Figure 6.2 illustrates the separation efficiency which
can be obtained from a pressure slurry-packed column of
20-μ silica at an operating pressure of 1500 psi. Columns of Corasil®-II provide similar performance.

Columns of alumina provide somewhat lower efficiencies
than corresponding silica columns. In our experience the
smallest particles of alumina which can be dry-packed to
give good columns are about 60 μ (i.e., larger than 200
mesh).

The water content of the adsorbent plays a critical
role in LSC. Some water must be added to the adsorbent
to increase linear capacity to a usable value and to maximize separation efficiency. Added water also decreases
the buildup of static charges during dry packing, thereby favoring denser, more efficient columns. The addition
of water to the adsorbent also decreases adsorbent-
catalyzed sample reactions (see later discussion) and
irreversible sample adsorption, thereby favoring complete
recovery of sample components. Since the water content
of the adsorbent markedly affects relative k' values and
band migration rates, water content must be held constant
for repeatable separation.

Maximum linear capacity in LSC (e.g., Figure 6.4)
usually occurs for 50-100% of a water monolayer, that is,
0.02-0.04 g of water per 100 m^2 of adsorbent surface.*

---

*Any reference to monolayers of added water simply indi-
cates the amount of water added per unit of adsorbent
surface (1 monolayer = 0.04 g of water per 100 m^2 of
surface).

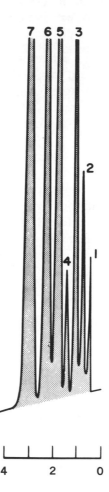

Figure 6.2. Separation of CCl₄ plus six aromatic
hydrocarbons on a 4-ft column of 20-μ silica;
elution with pentane at 100 atm.

With Porasil® A this means 6-12 g of water per 100 g of
dry adsorbent. Addition of the right amount of water can
raise adsorbent linear capacity by a factor of 5-100 (5),
relative to dry adsorbent.

For large-particle silicas (~ 100 µ), the addition of
1/3-1 1/2 monolayers of water increases the column plate
number by a factor of as much as 3 (other factors being
equal). In the case of Porasil® A an optimum water content of 8-15 g of water per 100 g of adsorbent is found.
This range of values overlaps the water content for maximum linear capacity and hence represents an optimum adsorbent water loading.

A number of different workers have assumed that the
manner in which water is added to the adsorbent affects
its final properties for use in LSC. In most cases, however, and especially with silica, any added water is rapidly equilibrated over the entire surface. Consequently
the manner in which water is added to the initial, dry
adsorbent is unimportant. The following procedure is
both convenient and satisfactory. The adsorbent is first
dried to remove all adsorbed water. With silica, this
can be accomplished by heating the adsorbent for 8-16 hr
at 110°C; alumina requires a temperature of 400°. The
dry adsorbent is allowed to cool in a closed metal container; then a weighed amount of adsorbent is transferred
to a closed glass container, with minimum exposure to the
atmosphere (the dry adsorbent rapidly adsorbs water from
the air). The required amount of water is added to the
glass container, and the stoppered container is shaken
until all lumps have disappeared. The resulting adsorbent is then allowed to stand for a day before being used.

The final adsorbent must be added to the column without allowing a change in adsorbent water content; that
is, there must be minimum contact of the adsorbent with
the atmosphere. In this regard it is helpful to use a
rubber-stoppered bottle for storage of the adsorbent,
with the upper half of a medicine dropper inserted into
the stopper (see Figure 6.3). A column can be loaded
directly from the bottle, by removing the bulb from the
dropper and placing the end of the dropper directly
against the mouth of the column.

For repeatable separations, where band migration
rates are to be held constant, adsorbent water content
must also be held constant, as mentioned earlier. This

Figure 6.3. Storage bottle for deactivated (standardized) adsorbent.

is achieved by adjusting the water content of the solvent (see the discussion in the following section). Slight differences in the starting adsorbent (i.e., as received), arising from batch-to-batch variations in a commercial adsorbent, can also be controlled by small adjustments in adsorbent water content. Thus a slightly more active adsorbent (larger retention volumes for some standard sample) can be compensated for by an increase in adsorbent water content, since k' and retention volume values decrease as adsorbent water content increases. In this way, sample retention volumes can be held within narrow limits for a given column or for different columns, as long as the starting adsorbent is approximately the same in surface area and particle diameter.

The size of the column is dictated by the length required for a certain efficiency, and in preparative separations by the quantity of adsorbent required for a given sample load (proportional to column volume). In the case of dry-packed columns, column diameters should be between 2 and 4 mm i.d. Columns of smaller diameter are difficult to pack (and generally give lower plate numbers), while columns of larger diameter are less efficient. The quantity of adsorbent required for a given sample size, or the amount of sample that can be placed on a given column, is defined by the adsorbent <u>linear capacity</u>, $\theta_{0.1}$: $\theta_{0.1}$ is the weight of sample per gram of adsorbent which causes a 10% reduction in the specific retention volume, V_R^o, of some standard compound, relative to the constant retention volume observed for smaller samples.*

This relationship is shown in Figure 6.4 for a typical case: dibenzyl as sample, pentane as solvent, and a

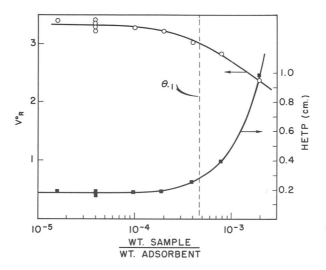

Figure 6.4. Dependence of solute retention volume (dibenzyl) and H on sample size (<u>13</u>). (Reprinted from Anal. Chem. by permission of the American Chemical Society.)

silica similar to Porasil® A. For small sample sizes

---

*The standard compound (plus solvent) must have a k' value greater than 1. The observed linear capacity of a column is larger for compounds with small k'.

the sample retention volume is independent of sample size
(and of the composition of the sample), but at some critical loading (the linear capacity, $\theta_{0.1}$) the retention
volume begins to decrease sharply with further increase
in sample size. At the same time we see in Figure 6.4
that separation efficiency (which is inversely proportional to H) rapidly declines in an overloaded column. Consequently, sample size should be less than the linear capacity of the column, both for constant sample migration
rates and for maximum separation efficiency. In the case
of preparative separations, sample sizes as large as $\theta_{0.1}$
can be tolerated.

The linear capacity of a given column can always be
determined for a particular sample by measuring retention
volumes as a function of sample size. With water-deactivated silicas, $\theta_{0.1}$ for a single, retained component (k'
>1) will be about 0.1-0.5 mg of sample per gram of ad-
sorbent. For mixtures in which no single component comprises most of the sample, $\theta_{0.1}$ will be somewhat larger.
Nonretained sample components do not affect linear capacity, and relatively large amounts of such components can
be added to the column without degrading the separation.
Thus the addition of a weak solvent to the sample will
not decrease the amount of original sample (i.e., free of
solvent) which can be added to the column. For a detailed discussion of linear capacity in LSC, see ref. 5.

2. Selecting a Solvent

The factors which must be considered in choosing a solvent for LSC were discussed in detail in Chapter 4. The
practical aspects of the solvent and the importance of
low-viscosity solvents need no further comment. Similarly, the selection of the right solvent strength, $\epsilon°$,
in LSC is straightforward, once an appropriate eluotropic
series has been defined. Table 6.4 provides several such
series for silica as adsorbent, and Table 6.5 gives similar data for alumina. Table 4.2 and Figure 4.2 of Chapter 4 provide additional data for alumina. For additional eluotropic series see ref. 5.

Table 6.4

Eluotropic Series for LSC on Silica

$\varepsilon°$	I	II	III
0.00	Pentane	Pentane	Pentane
0.05	4.2%v PrCl[a]/pentane	3%v CH_2Cl_2/pentane	4%v Benzene/pentane
0.10	10%v PrCl/pentane	7%v CH_2Cl_2/pentane	11%v Benzene/pentane
0.15	21%v PrCl/pentane	14%v CH_2Cl_2/pentane	26%v Benzene/pentane
0.20	4%v Ether/pentane	26%v CH_2Cl_2/pentane	4%v EtOAc[b]/pentane
0.25	11%v Ether/pentane	50%v CH_2Cl_2/pentane	11%v EtOAc/pentane
0.30	23%v Ether/pentane	82%v CH_2Cl_2/pentane	23%v EtOAc/pentane
0.35	56%v Ether/pentane	3%v Acetonitrile/benzene	56%v EtOAc/pentane
0.40	2%v Methanol/ether	11%v Acetonitrile/benzene	
0.45	4%v Methanol/ether	31%v Acetonitrile/benzene	
0.50	8%v Methanol/ether	Acetonitrile	
0.55	20%v Methanol/ether		
0.60	50%v Methanol/ether		

a   Isopropyl chloride.   b Ethyl acetate.

Table 6.5

Eluotropic Series for LSC on Alumina

$\epsilon°$	I	II	III
0.00	Pentane	Pentane	Pentane
0.05	8%v PrCl[a]/pentane	1.5%v CH_2Cl_2/pentane	4%v Ether/pentane
0.10	19%v PrCl/pentane	4%v CH_2Cl_2/pentane	9%v Ether/pentane
0.15	34%v PrCl/pentane	8%v CH_2Cl_2/pentane	15%v Ether/pentane
0.20	52%v PrCl/pentane	13%v CH_2Cl_2/pentane	25%v Ether/pentane
0.25	5%v MeOAc[b]/pentane	22%v CH_2Cl_2/pentane	38%v Ether/pentane
0.30	8%v MeOAc/pentane	34%v CH_2Cl_2/pentane	55%v Ether/pentane
0.35	13%v MeOAc/pentane	54%v CH_2Cl_2/pentane	81%v Ether/pentane
0.40	19%v MeOAc/pentane	84%v CH_2Cl_2/pentane	4%v Pyridine/pentane
0.45	29%v MeOAc/pentane	1%v MeCN[e]/ether	8%v Pyridine/pentane
0.50	44%v MeOAc/pentane	5%v MeCN/ether	13%v Pyridine/pentane
0.55	65%v MeOAc/pentane	14%v MeCN/ether	20%v Pyridine/pentane
0.60	5%v PrOH[c]/ether	36%v MeCN/ether	32%v Pyridine/pentane
0.65	10%v PrOH/ether	MeCN	
0.70	20%v PrOH/ether		
0.75	3%v MeOH[d]/ether		
0.80	7%v MeOH/ether		
0.85	17%v MeOH/ether		
0.90	40%v MeOH/ether		

a Isopropyl chloride. b Methyl acetate. c Isopropanol. d Methanol.
e Acetonitrile.

For gradient elution or stepwise elution we need a sequence of solvents of continually increasing strength. The selection of an optimum solvent sequence in this connection has been referred to in Chapter 4 and is discussed in detail in ref. 10 for the specific case of LSC.

It should be noted that a change in solvent from one composition to the next in the series of Tables 6.4 and 6.5 (i.e., a change in $\epsilon°$ by 0.05 unit) means a change in k' by a factor of 2-4. For finer adjustments of sample k' values, the data of Tables 6.4 and 6.5 can be interpolated. For example, an $\epsilon°$ value of 0.22 on silica is provided by 7%v ether/pentane, 36%v CH_2Cl_2/pentane, or 7%v ethyl acetate/pentane.

The variation of the solvent in a given case for improved separation selectivity is governed by the rules discussed in Chapter 4. Solvent strength, $\epsilon°$, must be held approximately constant while solvent composition is varied. This is most conveniently done with solvent binaries (or multicomponent solvents) and is most effective when the nature of the strong solvent component is changed in the solvent mixture. For example, if 10% (v/v) ether/pentane provides k' values of the right magnitude, but two bands are unresolved because α is close to 1, some other moderately strong solvent can be substituted for the ether in the hope of altering α (e.g., 10%v ethyl acetate/pentane; see Table 6.4). Replacing the pentane in this solvent mixture by another weak solvent, such as hexane or benzene, would be unlikely to have as large an effect on α. Strong solvent selectivity in LSC usually arises from solute-solvent-adsorbent interactions, and it is the strong solvent component which largely comprises solvent in the adsorbed phase.

During repeated separations on a water-deactivated adsorbent, a dry solvent will gradually remove water from the adsorbent and change its chromatographic properties (for the worse). This can be avoided by adding just enough water to the solvent so that solvent and adsorbent are in thermodynamic equilibrium with respect to water. Thus continued flow of solvent through the column will not change the water content of the adsorbent. The

principle of adjusting the water content of the solvent
is easy to understand: the thermodynamic activities of
the water in the solvent and on the adsorbent must be
made equal. In practice, however, certain technical difficulties arise which we will now examine. First, there
is the problem of deciding how much water is to be added
to the solvent. Second, the addition of water to water-
immiscible solvents poses certain problems. Finally, we
sometimes want to adjust the water content of the adsorbent after the column is packed, and the rate of equilibration of solvent and adsorbent with respect to water is
in some cases quite slow.

The determination of how much water is to be added to
the solvent depends on whether the solvent has a relatively low or high capacity (i.e., solubility) for water.
Weak (nonpolar) solvents are usually saturated by relatively low concentrations of added water (e.g., 20-100
ppm), while strong (polar) solvents are in some cases
completely water miscible. The effect of water content
on adsorbent properties becomes less pronounced as solvent strength increases, so that precise adjustment of
solvent water content is most important for relatively
weak solvent systems. For strong solvents it is convenient to add varying amounts of water and then determine
the retention volume of some solute (for which $1 < k' < 5$)
in repeat runs on a fresh column of adsorbent. For example, the column is prewet with solvent, the sample is
injected, and the first retention volume is determined.
This is followed by a second injection of solute and the
determination of a second value for the retention volume.
If the retention volume changes between the first and the
second injections, the water content of the solvent is
not in equilibrium with that of the adsorbent. A lower-
ing of the second retention volume means too wet a solvent, while an increase in the second retention volume
means too dry a solvent. By trial and error it is possible to determine the right solvent water content, that
is, the water content which leads to identical retention
volumes in the first and the second injection.

It is difficult to add controlled amounts of water to
nonpolar solvents because of the low rate of solution of

water. Another problem is entrainment of small amounts
of undissolved water as tiny droplets which escape notice
and subsequently deactivate the adsorbent at the column
inlet. These problems can be overcome by saturating one
batch of solvent with water (e.g., by percolating solvent
through a column of 30% (by weight) water on Davison code
62 silica) and then blending the wet solvent with dry
solvent to give a certain per cent water saturation in
the final solvent [e.g., (equal parts wet plus dry solvent) = 50% saturation]. The correct degree of water
saturation can then be determined by trial and error,
just as in the case of strong solvents. A 200 m^2/g alumina with 4% added water will be in equilibrium with 25%
water-saturated solvent, while 10% water on Porasil® A
will be in equilibrium with about 50% water-saturated
solvent.

Sometimes it is necessary to adjust the water content
of the adsorbent after it has been packed into the column. For example, some inadvertent change in the adsorbent water content may have occurred. In such cases it
is sufficient to flow enough solvent of the right water
content through the column. When the retention volume of
some standard sample shows no change with further solvent
flow, the column will have equilibrated with the solvent.
It should be noted that nonpolar solvents require a long
time to reach equilibrium with the adsorbent (see, e.g.,
ref. 1), because of their limited solubility for water.
In some cases it may be necessary to flow several hundred
column volumes of solvent through the column before equilibrium is reached. In testing a nonpolar solvent to see
whether its water content is in equilibrium with the adsorbent (as above), 10-20 column volumes of solvent
should flow through the column between successive injections.

3. Column Temperature Effects

A deliberate change in the temperature of the column
is seldom worthwhile in LSC. Special separation effects
due to temperature variation usually can be duplicated
more easily by changing the adsorbent water content or

the solvent. The only important exception occurs in the
preparative separation of difficultly soluble samples,
where an increase in column temperature may be the only
way of dissolving sufficient sample. The general effect
of varying temperature in LSC is a decrease in k' values
as temperature increases. Usually k' decreases by about
1% per 1°C rise in temperature, a change which seldom
leads to any problem in unthermostatted columns. For
these reasons most LSC separations are carried out at
room temperature, without any attention to temperature
control of the column.

D. SPECIAL PROBLEMS IN LIQUID-SOLID CHROMATOGRAPHY

In most cases the application of LSC to a given separation problem will prove to be straightforward and
trouble-free (at least for the sample types which are
best suited to LSC; see Section A of this chapter). However, we should note three rather general problems which
occasionally arise in LSC:

- Alteration or loss of sample during separation.
- Difficulty in reproducing adsorbents.
- Unstable column operation.

Alteration or loss of sample during separation can
arise from adsorbent-catalyzed reaction of the sample or
irreversible adsorption of the sample. Reaction of the
sample during LSC has been a common complaint during the
past 40 years. The commonly used adsorbents are often
acidic or basic, and samples sensitive to pH can be altered on such adsorbents. Sample oxidation is usually
increased in the presence of the adsorbent. The solution
to these problems is to avoid acidic adsorbents with
acid-sensitive samples, and basic adsorbents with base-
sensitive samples, and to take precautions to exclude
oxygen from the separation system (particularly for
easily oxidized compounds).

Silica is weakly acidic (pH ~4) as a result of the
surface -OH groups, and commercial products sometimes
contain traces of strong acid left over from the gelation
step. In the latter case it is possible to remove this

acid by water washing (to neutral pH). Samples which are
extremely sensitive to weak acid can be separated on silica by incorporating a basic buffer into the solvent system. Alumina and magnesia are examples of basic adsorbents, and ordinary alumina has a surface pH of about 12.
Base-sensitive samples cannot be separated on alumina unless the adsorbent is first acid-treated.

Oxidation during LSC separation can be reduced by
flushing the column and all solvent with nitrogen (to
remove oxygen) and by adding an antioxidant to the solvent. The antioxidant 2,6-di-t-butyl-p-cresol is particularly useful in this regard. Addition of 0.005% of this
compound (DBPC-BHT) to the solvent is effective in sup-
pressing oxidation, has no effect on solvent strength (at
this concentration), and is easily removed from separated
fractions by evaporation at room temperature or slightly
higher temperatures (12).

The addition of water to the adsorbent (which is useful for other reasons) tends to suppress all types of
sample reactions. The possibility of solvent reactions
during separation should also be kept in mind. For example, acetone cannot be used on alumina as a solvent
for preparative separations. In the presence of this
basic adsorbent acetone polymerizes to diacetone alcohol
and higher-boiling products, and these then contaminate
the separated sample fractions.

The problem of sample reaction during separation is
much less severe in modern, high-efficiency LSC than in
previous applications of this technique. First, the
separation times now required are often 1 or 2 orders of
magnitude less than in classical LSC separations, and
this means a greatly reduced reaction time during separation. Consequently, the extent of reaction will be very
much less, even where sample reaction is possible in an
LSC system. Second, in comparison to TLC, it is much
easier to protect the sample from air and light during
separation on columns.

Irreversible retention of part or all of the sample
during LSC separation is occasionally encountered, particularly for very polar samples. In some cases this

problem can be overcome by water deactivation of the adsorbent or changes in the solvent system. Florisil® (an alumina-silica adsorbent) has often been noted to give irreversible sample adsorption, even for nonpolar compounds. Addition of water to Florisil® has only a minor effect on irreversible adsorption and sample loss.

The general difficulty in reproducing adsorbent batches has been cited frequently in the past. Any given type of adsorbent, such as silica, can show wide variations in its chromatographic behavior as a result of corresponding variations in adsorbent surface area and pore structure, or the nature of the adsorbent (e.g., crystallinity or composition). In our experience, particularly today, a given commercial product such as Porasil® A will show minor variations in its properties from batch to batch. However, these can largely be controlled by small variations in the amount of water added to the adsorbent (see the previous discussion).

Unstable column operation (i.e., variation in the performance of a column during use) can occur in LSC as a result of adsorbent deactivation or a change in adsorbent water content. Adsorbent deactivation can occur from the presence in the sample (or solvent) of strongly adsorbing minor components. With repeated column use these gradually build up on the adsorbent, thereby deactivating the surface and leading to reduced retention volumes and decreased column plate numbers. The best solution to this problem is the use of a guard column: a short length of column of the same composition as the main column. The guard column is connected between the sample inlet and the main column, and it is periodically replaced. In this way any strongly adsorbed components are collected on the guard column and discarded before they can leak through to the main column. If the main column becomes deactivated by these strongly adsorbed components, it is usually better to discard the column and pack a new one. Alternatively, if the main column represents an unusual investment of time or money, column regeneration can be used. A solvent of very high strength (methanol, pyridine, dimethyl sulfoxide, etc.) can be used to wash the

strongly held components from the column. The strong solvent is then flushed from the column with a somewhat weaker solvent (e.g., $\epsilon°$ 0.2 unit smaller), this solvent is
removed with a still weaker solvent, and the process is
repeated until the original solvent has been reverted to
(i.e., all strong solvents have been washed from the column). Then it is only necessary to restore the adsorbent
to its correct water content as described above.

Unstable column operation as a result of changes in
adsorbent water content is easy to understand. In principle, it should suffice to control the water content of
the solvent closely. In practice, this can prove somewhat difficult, particularly for nonpolar solvents. The
problem is further complicated by the slow response of
the column to any deliberate (or unintentional) variation
in solvent water content. If a change in solvent water
content is indicated to restore the original water content
of the adsorbent, it is advisable to make only small
changes at one time. Sufficient time for column equilibration must then be allowed. Further adjustments in solvent water content (after equilibration) can be made, if
necessary.

E. APPLICATION OF LIQUID-SOLID CHROMATOGRAPHY
TO SOME PRACTICAL PROBLEMS

As noted in Section A, practical applications of high-
efficiency (i.e., Section A, "modern") LSC are still relatively few in number. This is particularly true for
reported applications. It will be instructive to examine
two examples of this type, mainly for the insight they
provide into how different LSC techniques can be used in
the solution of a given problem.

The first example was reported recently (11). Dilute
solutions of quinoline in dodecane had been hydrogenated
catalytically, and it was required to determine what products were produced. Major interest centered on the aromatic, nitrogen-containing products. The total concentration of nitrogen in the hydrogenated products ranged from
1000 ppm down to less than 1 ppm. Figure 6.5 shows the
principal products of interest, although they were not

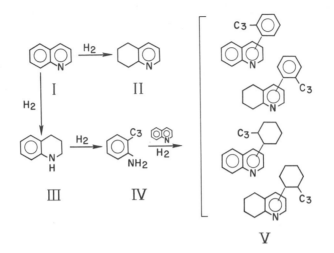

Figure 6.5. Products from the hydrogenation of
quinoline in dilute solution.

known at the start of the project. The first step was the
qualitative analysis of a typical product sample to determine what compounds were present. Gradient elution is
well suited to initial, exploratory separation for determining the compositional range of a sample. Figure 6.6
shows the chromatogram obtained from the gradient elution
(LSC) separation of a typical product sample. The major
components (numbered) fall in the middle of the chromatogram, with minor amounts of hydrocarbons (i.e., nitrogen-
free compounds) at the beginning, and minor amounts of
more polar compounds at the end. The major bands shown
in Figure 6.6 were readily isolated by scaling up this
gradient elution separation. The identity of each isolated band was established by spectroscopic analysis, as
indicated in Figure 6.6. The later-eluted, minor components were also characterized but were found to be highly
rearranged compounds of little interest.

Band V-b in Figure 6.6 occurs as an unresolved shoulder. The resolution of the LSC system was increased by

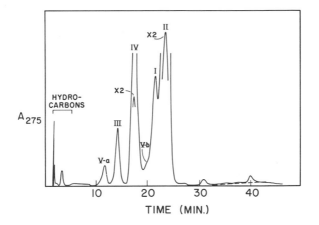

Figure 6.6. Gradient elution separation on alumina of hydrogenated quinoline sample (11). (Reprinted from J. Chromatog. Sci. by permission of
the editor.)

using a single solvent* of optimum strength, a longer
column, and a slower solvent flow rate. The resulting
separation of band V-b is shown in Figure 6.7. Three
distinct bands are now observed between bands IV and I.

After the nature of the products formed from the
hydrogenation of quinoline had been established, it was
desired to analyze several hundred samples for the compounds of interest: I, II, III, IV, V-a, and V-b. A
detailed breakdown of band V-b into its three components
was considered unnecessary. As a result it was possible
to use a single-solvent system (since bands I-V elute
within a narrow range) of lower resolution than that
shown in Figure 6.7. This simplified the routine analysis of these samples and reduced the necessary time

---

*In gradient elution the elution of a given peak corre-
sponds to a given solvent strength (for the column effluent) at the time of elution. Optimized single-solvent
separation of the same band will generally require a solvent of slightly greater strength.

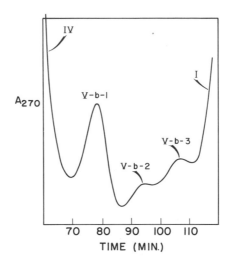

A_{270}

70 80 90 100 110
TIME (MIN.)

Figure 6.7. High-resolution separation of Band V-b
(<u>11</u>). (Reprinted from J. Chromatog. Sci. by permission of the editor.)

per sample. The resulting separation is shown in Figure
6.8. The essentially complete resolution of all bands of
interest permits their easy quantitation.

A second (unpublished) application of LSC is concerned
with the determination of antioxidants present in unknown
samples (e.g., competitor products). Figures 6.9 and
6.10 illustrate the separation of some typical antioxidants on two different LSC columns. Nonpolar antioxidants
are well resolved in the system of Figure 6.9, but polar
antioxidants require an excessive elution time. Similarly, polar antioxidants are well resolved on the Corasil®-
II column of Figure 6.10, but nonpolar antioxidants are
rapidly eluted as an unresolved band. The good characteristics of both of these LSC systems can be combined
into a single separation, using the technique of coupled
columns described in Chapter 4.

In the present case a variation of this technique is
used, based on the column arrangement shown in Figure
6.11. By using constant pressure, flow programming (see

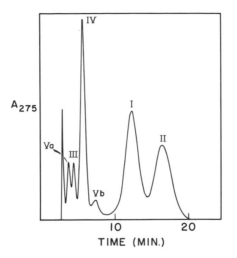

Figure 6.8. Medium-resolution separation of hydrogenated quinoline sample by normal elution from alumina (11). (Reprinted from J. Chromatog. Sci. by permission of the editor.)

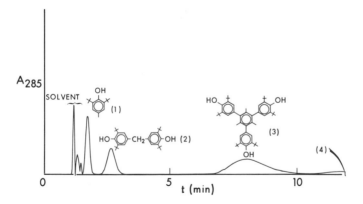

Figure 6.9. Separation of antioxidants on 100 cm Porasil® A with 17%v CH_2Cl_2/pentane as solvent.

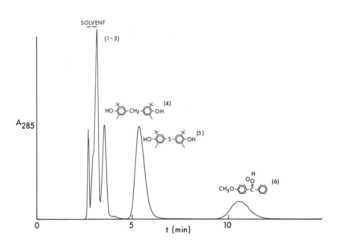

Figure 6.10. Separation of antioxidants on 150 cm Corasil® II with 17%v CH2Cl2/pentane as solvent.

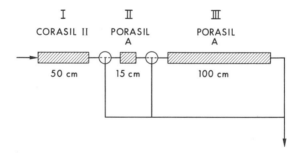

Figure 6.11. Column arrangement for coupled-column flow-programmed separations.

Chapter 4) is combined with coupled-column operation. The sample is injected and eluted through columns I, II, and III (in sequence). Just before nonretained peaks reach the end of column III (100 cm Porasil® A), the first three-way valve is turned so that direct elution from column I (Corasil®-II) occurs into the detector. Rapid elution of polar antioxidants from column I now occurs, because of the low capacity of Corasil®-II and the rapid flow through this column. When all components

of the sample have been eluted from column I, the two
three-way valves are changed to allow elution through
columns I and II. Components initially held on column
II (antioxidants of intermediate polarity) are rapidly
eluted from this column, because of its short length and
relatively high solvent flow rate. Once elution from
column II is completed, the three-way valves are set to
allow flow through all three columns in sequence. Components held initially on column III (weakly polar antioxidants and other compounds) are now slowly eluted from
this column, because of its length and the slow solvent
flow rate. Under these conditions weakly retained compounds are well resolved.

The resulting separation of the antioxidant mixture
of Figures 6.9 and 6.10 is shown in Figure 6.12. Adequate separation of all six compounds is observed, within

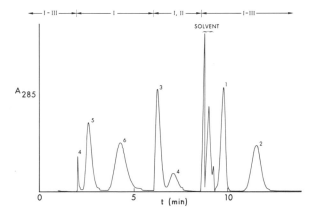

Figure 6.12. Antioxidant separation (mixture of
Figures 6.9 and 6.10) by the system of Figure
6.11; coupled columns, 17%v CH_2Cl_2/pentane

a reasonable time. The similar separation of an actual
polymer extract is shown in Figure 6.13. Here at least
nine aromatic compounds are seen, two of which were subsequently identified as antioxidants.

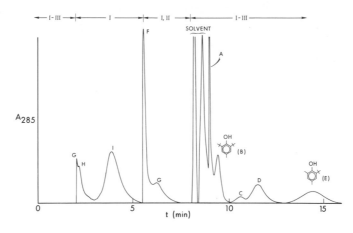

Figure 6.13. Polymer extract separation by the system of Figure 6.11; coupled columns, 17%v CH_2Cl_2/pentane.

References

1. L. R. Snyder, J. Chromatog. Sci., 7, 352 (1969).
2. E. Stahl, "Thin-Layer Chromatography: A Laboratory Handbook," 2nd ed., Academic, New York, 1969.
3. L. J. Morris and B. W. Nichols, in "Chromatography," 2nd ed., E. Heftmann, ed., Reinhold, New York, 1967, p. 466.
4. L. R. Snyder, Anal. Chem., 39, 698 (1967).
5. L. R. Snyder, "Principles of Adsorption Chromatography," Marcel Dekker, New York, 1968.
6. L. R. Snyder, J. Chromatog., 8, 319 (1962).
7. L. R. Snyder, J. Chromatog., 17, 73 (1965).
8. L. R. Snyder, J. Chromatog., 20, 463 (1965).
9. L. R. Snyder, Accounts Chem. Res., 3, 290 (1970).
10. L. R. Snyder, J. Chromatog. Sci., 7, 195 (1969).
11. L. R. Snyder, J. Chromatog. Sci., 7, 595 (1969).
12. J. J. Wrenn and A. D. Szczepanowsa, J. Chromatog., 14, 405 (1964).

CHAPTER 7

The Practice of Gel Permeation Chromatography

Karl J. Bombaugh

A. THEORY OF GEL PERMEATION CHROMATOGRAPHY

Gel permeation chromatography (1) is a form of liquid chromatography in which solute molecules are retarded as a result of their permeation into solvent-filled pores in column packing. Larger molecules excluded from all or a portion of the pores by virtue of their physical size elute from the column before the small molecules, thereby providing a separation based on molecular size in solution. The separation may be described by the classical chromatography equation

$$V_R = V_M + K_o V_s \qquad (7.1)$$

where V_M = the column void, or transport volume (interstitial volume),

V_s = the volume of solvent imbibed in the pores of the column packing,

$K_o = K(V_{fs}/V_s)$ = volume fraction of the stationary phase available to the solvent molecules divided by the volume of the stationary phase,

K = concentration in the stationary phase (2) divided by concentration in the moving phase.

The separation is described with the aid of Figure 7.1.

1. The Separation Mechanism, Using the Steric Exclusion Model*

When a population of molecular species with a size distribution as shown in Figure 7.1 is injected as a solute

---

*The steric exclusion model proposed by Flodin (see below) is used since it is most readily adaptable to generally accepted chromatography theory. Furthermore, this mechanism is adequate to describe the principles of GPC in language familiar to the liquid chromatographer. A

237

number of mechanisms have been proposed to describe the size-separating process; these generally fall into three categories:

(1) Steric Exclusion

    (a) P. Flodin, thesis, University of Uppsala, Uppsala, Sweden (1962).

    (b) (Secondary exclusion) K. H. Altgelt. Preprints Div. Petroleum Chem. Am. Chem. Soc., 15 (2), A115 (February 1970).

(2) Restricted Diffusion

    (a) G. K. Ackers, Biochem., 3 (5), 723 (1964).

    (b) W. W. Yau and D. P. Malone, J. Polymer Sci., Part B-5, 663 (1967).

    (c) E. Z. DiMarzio and C. M. Guttman, J. Polymer Sci., 7, 267 (1969).

(3) Thermodynamic Theories

    (a) A. J. Devries, M. LePage, R. Beau, and C. I. Guillemin, Anal. Chem., 39, 935 (1967).

    (b) M. J. R. Cantow, R. S. Porter, and J. F. Johnson, J. Polymer Sci., Part A-1, 5, 987 (1967).

    (c) E. F. Casassa, J. Polymer Sci., Part B-5, 773 (1967).

A discussion of these mechanisms is beyond the scope and, indeed, the need of this text. However, an excellent summary of the principles involved, including an extensive bibliography, is contained in a review article by D. L. Bly in Science, 168, No. 3931, 527-533 (1970).

---

band into a gel column, molecules tend to diffuse away from the concentration center, which directs them into the solvent trapped in the gel pore. When the solvent inside the pore is the same as the solvent outside of the pore (K = 1), the permeation is purely a function of entropy. Under static conditions, permeation proceeds until equilibrium is reached. However, with carrier flow,

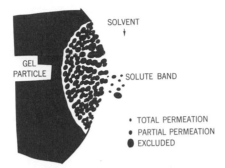

<u>Figure 7.1.</u> Model of the permeation process

the solute band is moved along the column. Under these
conditions, permeation occurs while the concentration outside the gel exceeds the concentration inside the gel.
When the solute band is moved from a given zone, the concentration inside the gel exceeds the concentration outside. The molecules then diffuse out into the carrier
stream, where the process is repeated cyclically throughout the length of the column.

By this process, all molecules, given adequate time,
will permeate all of the solvent available within the gel.
As depicted in Figure 7.2, molecules of size A or larger
excluded by the gel pore are unretained at V_M. Molecules
of size B or smaller permeate into all of the pores and
elute at total permeation volume, V_t. Molecules ranging
in size between A and B permeate into, and elute in accordance with, the volume fraction of solvent available
to them. It is evident that the volume fraction available cannot exceed the total volume, V_s. Consequently,
K_0 should not exceed 1. When K_0 is greater than 1, K
must also be greater than 1, and the separation is not
purely a permeation process. In this treatment, it will
be assumed that $K_0 = 1$, where the separation is purely
an entropy-exclusion process.

The limiting value of K_0 produces some unique and desirable properties, as well as some limitations to

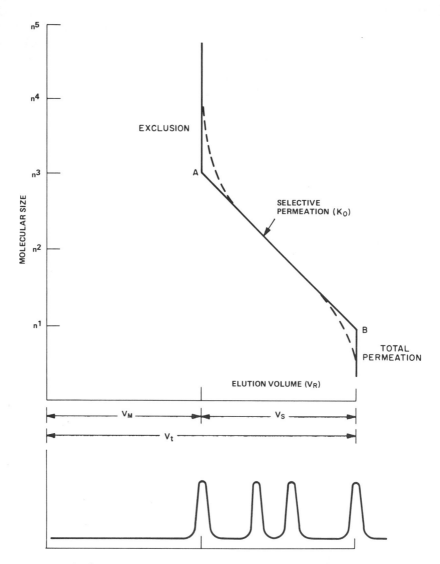

Figure 7.2. Descriptive calibration curve for the permeation process.

exclusion chromatography. Gel permeation chromatography is ideal chromatography. Isotherms are linear, that is, concentration in the bulk stationary phase is always directly proportional to concentration in the moving phase, and peaks are Gaussian. Since K_O has a maximum of 1, all components must elute within a fixed volume, V_t. No late peaks are encountered. Sample injection may be sequenced without waiting for the last peak to elute. Automatic sample injection is practicable, even with unknown samples.

2. What Is Measured?

When K = 1, elution volume is a measure of a physical property of the species, that is, its physical size in solution; the size of the species in solution is approximately proportional to molecular weight, which may be related to elution volume by a calibration curve. The calibration curve shown in Figure 7.2 is ideal and is drawn as three straight lines with sharp intersections (3). In practice, however, the calibration curve is rounded and continuous, as represented by the dashed line. This occurs because the gel pores are not uniform in size, but rather occur with a distribution of sizes. Furthermore, the molecules in solution are in constant motion and can assume a variety of conformations having a broad range of molecular dimensions. The resulting calibration then represents a statistical average of the molecular sizes and pore distribution in the column. When the pore distribution range is wide, the curve will be steep. A column of fixed length will then provide less resolution but will offer a wider linear range, $(\Delta \log M_w)/\Delta V_R$, where M_w is the molecular weight of the species. When pores are all one size, the curve will be flat, offering high resolution between species whose size is within the flat range of the curve.

A typical GPC chromatogram, as used to determine the molecular weights of high polymers, is shown in Figure 7.3. This plot represents the molecular weight distribution of polyvinyl chloride, containing perhaps more than 10,000 species, which differ from each other incrementally by but a few molecular weight units. To the analytical chromatographer, the absence of individual peaks may suppress interest in the technique. This would be a

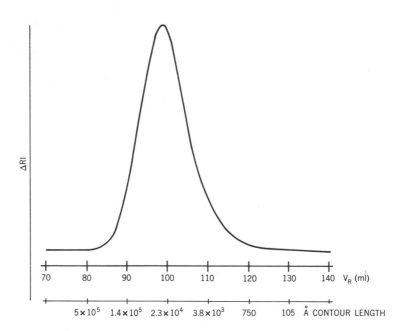

<u>Figure 7.3.</u> Typical GPC chromatogram of a poly-
mer distribution. <u>Conditions</u> - column: Styragel®
10^6, 10^5, 10^4, 10^3 Å porosity; flow: 1 ml/min;
solvent: tetrahydrofuran; temp.: 22°C; sample:
5 mg polyvinyl chloride.

costly error, however, since separation of discrete species by GPC is entirely practical (<u>2</u>-<u>11</u>).

The chromatogram in Figure 7.4 also represents a
molecular weight distribution. In this case, the system
resolution is increased to separate discrete species (<u>8</u>).
The mechanism in each case is the same; only the conditions differ. It is evident from Figure 7.4 that exclusion chromatography is capable of separating molecular
species according to size, thereby providing a useful,
high-resolution analytical technique.

3. <u>Resolution</u>

Resolution in GPC, as in other forms of chromatography, is described by the resolution equation:

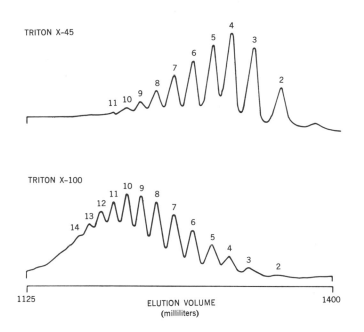

<u>Figure 7.4</u>. High-resolution GPC separation of
surfactants (<u>8</u>) (nonylphenol-ethylene oxide adducts). <u>Conditions</u> - column: 160-ft 3/8-in. of
500 Å Styragel®; solvent: tetrahydrofuran; sample:
2.5 mg; flow: 0.4 ml/min. Triton® X-45 is a registered trademark of Rohm & Haas Co.; numbers 2-4 =
moles ethylene oxide/mole nonylphenol.

$$R_s = \frac{1}{4}\left(\frac{\alpha}{\alpha - 1}\right)\left(\frac{k'}{1 + k'}\right)(\sqrt{N}) = \frac{1}{4}\left(\frac{\alpha}{\alpha - 1}\right)(\sqrt{N_{eff}}) \quad (7.2)$$

This equation expresses resolution in terms of selectiv-
ity (α), capacity (k'), and efficiency (N) (<u>12</u>-<u>15</u>). Although resolution was discussed in Chapter 1, it is de-
sirable to describe R_s as it applies specifically to gel
permeation chromatography.

 <u>Selectivity</u>. The selectivity term in GPC relates
purely to the difference in molecular size. Consequently,
improving resolution by increasing the specificity term

is accomplished only by optimization of the pore size and
the pore size distribution of the gel, which increases
the X axis of the calibration curve relative to the Y
axis. In simple terms, to increase the selectivity term,
one should use gels with uniform porosity in the size
range to match the solute, so as to produce a flat calibration curve.

Efficiency. The efficiency term in GPC, as in any
form of chromatography, is an inverse measure of the random dispersion (15). The uniform spherical particles
produced by suspension polymerization produce highly efficient columns. Plate numbers of 1000 plates per foot
(measured at linear velocities of 0.1 cm/sec and at K =
1) are common in GPC, compared to 200 ppf in other forms
of liquid chromatography. One difference between GPC and
other forms of liquid chromatography, however, is that
peak widths (w) are equal at all values of K_O (8). This
is in contrast to sorption chromatography, where peak
widths increase as K increases.

Since the peak widths of small molecules are equal,
plate number increases with increasing values of K_O, as
shown in Tables 7.1a and 7.1b (8). Solutes with K_O values from 0 to 1 all showed w $\cong$ 14 ml, yielding N values
ranging from 100,000 to 180,000 plates (arithmetically,
as V_R increases at constant w, N must increase). As a
convenience, however, one may measure w at V_t and apply
the result to any value of K_O to calculate the N or N_{eff}
required for a separation (8).

The uniform peak width is probably a coincidence resulting from compensating interactions. Since the diffusion coefficient decreases with increasing molecular size,
the band spreading resulting from slowed diffusion is
exactly compensated for by the reduced time the material
spends in the stationary solvent. When a particular solute in a mixture has a lower diffusion coefficient than
another of the same size, it will show a wider peak width.
[For example, o-dichlorobenzene produces a wider peak
than either decane or acetonitrile (2).]

Capacity. The capacity term in GPC has, unfortunately, been neglected by most practitioners, yet this factor

Table 7.1a

GPC Separation Data for Triglycerides (8)

Component	Carbon Number	Molecular Weight M_w	Retention Volume V_R, ml	Peak Width w, ml	Theor. Plates N	Relative Retention α	Required Theor. Plates N req	Reso- lution σ
Polystyrene		867,000	873	14.50				
Triarachidin	63	976	1132	14.30	100,240	1.0134	205,900	4.2
Tristearin	57	892	1147	14.60	98,750	1.0141	186,220	4.6
Tripalmatin	51	807	1163	13.35	121,460	1.0165	136,630	5.6
Trimyristin	45	723	1183	14.30	109,380	1.0178	117,700	6.1
Trilaurin	39	639	1203	13.10	135,010	1.0202	91,830	7.3
Tricaprin	33	555	1228	13.55	131,350			
O-Dichlorobenzene		147	1579	14.85	180,840			

Table 7.1b

Effective Plate Calculations for Triglycerides (8)

Component	Retention Volume VR, ml	VR-VM, ml	Peak Width w, ml	Effect. Plates Neff	Theor. Plates N	Capacity Ratio k'	Relative Retention α	Effect. Plates Neff	Theor. Plates Nreq	Resolution σ
Polystyrene	873.2		14.50							
Triarachidin	1131.8	258.6	14.30	5,234	100,240	0.296	1.0586	11,630	206,760	4.2
Tristearin	1147.0	273.8	14.60	5,627	98,750	0.313	1.0589	11,630	187,320	4.6
Tripalmitin	1163.2	290.0	13.4	7,548	121,460	0.332	1.0662	9,340	136,340	5.6
Trimyristin	1182.3	309.1	14.3	7,478	109,380	0.354	1.0679	9,130	120,520	6.1
Trilaurin	1203.4	330.2	13.1	10,162	135,010	0.378	1.0738	7,580	91,000	7.3
Tricaprin	1227.7	354.5	13.6	10,952	131,350	0.405				
o-Dichlorobenzene	1578.08	705.6	14.8	36,118	180,840	0.808				

is of prime importance to the technique. The capacity ratio (k') is a measure of the time (t) that the solute spends in the stationary phase, relative to the time in the moving phase. The term k' is relatively small in GPC because K_O never exceeds 1. To increase k' one must increase V_s relative to V_M. Consequently, the V_s/V_M ratio (i.e., k' at $K_O = 1$) may be used to evaluate gel capacity. Although capacity values range between 0.5 and 2, the most widely used rigid gels show capacity values near 1.

From the resolution equation, it is evident that by doubling the pore volume (V_s/V_M) of a gel, the plate requirement, or column length, may be reduced by one-fourth. The principle is illustrated in Figure 7.5 for a separation in which Heitz, et al. (16) used highly swollen gels with a large pore volume, on a 2 meter column, producing 25,000 theoretical plates. The soft gels afforded a V_s/V_M of 2.2. In contrast, the separation shown in Figure 7.4, for which a gel with a V_s/V_M of 0.8 was used, required a 160-ft column, generating 180,000 theoretical plates at V_t to accomplish a similar separation. It is evident, therefore, that V_s/V_M is a valid measure of the separating capacity of a GPC packing.

Particular attention must be called to the relevance of the use of effective plates (N_{eff}) in GPC because of the small capacity factors available. This is clearly demonstrated by the data in Table 7.1b. By comparing the data in the theoretical plate (N) column with those in the effective plate column (N_{eff}), the contribution of the large column dead volume relative to the working volume becomes evident. At $k' = 0.8$, 180,000 theoretical plates represent 36,000 effective plates. An even greater effect is shown for materials which spend less time in the stationary phase. At $k' = 0.3$, 100,000 theoretical plates represent only 5000 effective plates when correction is made for V_M.

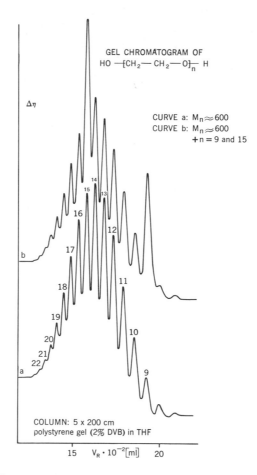

Figure 7.5. High-resolution separation with soft gels [per Heitz et al. (16)]. Conditions - Column: 5 x 200 cm of polystyrene gel, 2 cross-linkers in tetrahydrofuran; sample: polyethylene glycol; curve a: $\simeq M_n$ 600, curve b: $\simeq M_n$ 600, + n = 9 and 15.

B. PRACTICAL ASPECTS OF GEL PERMEATION CHROMATOGRAPHY

1. Column Packing Materials

Numerous packings are available for gel permeation chromatography. They are listed by chemical type in

Table 7.2, where they are considered as being soft, semirigid, or rigid. The classification is significant,
since both gel performance and operating technique are
related to this property.

Soft Gels. Soft gels are lightly cross-linked structures capable of imbibing large quantities of solvent
into their pores. These materials swell to many times
their dry volume and gain their porosity in proportion
to the volume of solvent imbibed. Soft gels have been
used primarily with aqueous solvents; the technique is
then known as gel filtration chromatography.

Soft gels offer a trade of capacity for permeability.
Because of their softness, these materials deform in the
inertial field of moving solvent, thereby increasing column pressure drop. As the gels deform, at high solvent
velocity, the bed compresses, producing voids and hence
destroying the column. At high flow velocity, soft gels
have been known not only to compress, but also to extrude
through the screens in the column end fittings. Furthermore, since pore size is proportional to swelling factor,
soft gels become more fragile at increased porosity. For
this reason, they tend to be more usable with small molecules and must always be employed at low flow velocities.

When used at low flow velocities, the soft gels offer
high efficiency and high capacity. With lightly cross-
linked gels, V_S/V_M ratios approaching 3 have been reported. With equal efficiency, such a gel would afford a
separation in a column one-ninth the length of a column
that offered a V_S/V_M ratio of 1.

Semirigid Gels. Semirigid gels, produced by "pearl"
polymerization (1), provide a compromise of high permeability and average capacity over a wide range of pore
sizes. In contrast to soft gels, which swell to many
times their dry volume, semirigid gels swell to only 1.1-
1.8 times their dry volume. As open-caged spheres, they
withstand high pressures without deformation over their
entire porosity range, from small molecules to macromolecules in excess of 10^6 molecular weight. Capacity (V_S/V_M)
ranges from 0.8 to 1.2. Column efficiencies in the range
of 1000 ppf are routinely afforded by the uniform spherical particles.

Table 7.2

Column Packing Materials for Exclusion Chromatography

| Soft | | Semirigid | | Rigid | |
Organic	Aqueous	Organic	Aqueous	Organic	Aqueous
2% Cross-linked polystyrene	Polydextran	Polystyrene	Polystyrene	Glass	Glass
	Polyacrylamide	Polyvinyl acetate	Sulfonated	Silica	Silica
Chemically modified polydextran	Agarose	Polymethyl meth-acrylate	Merrifield derivative	Modified silica	Modified silica
Rubber	Starch				

High-speed liquid chromatography in the modern sense is possible with semirigid gels, as illustrated by the chromatogram shown in Figure 7.6. The 80-sec separation of three components to base line was made at a solvent velocity of 2 cm/sec (2). The separation of Triton® shown in

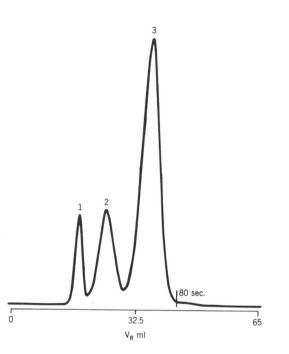

Figure 7.6. Fast GPC chromatogram (2). Conditions - column: 3 ft x 3/8 in. o.d. of 100 Å Poragel®; flow: 32.5 ml/min, v = 2.82 cm/sec; solvent: toluene; temp.: ambient. Components - 1: 7.5 mg polypropylene glycol 4000; 2: 7.5 mg hexadecane; 3: 15.0 mg acetonitrile.

Figure 7.4 was made on a column 160 ft in length that was operated for more than 6 months at 1000 psi, using flow rates of 0.1 cm/sec, without any evidence of deterioration in column performance.

Currently available semirigid gels are applicable primarily with organic solvents; the technique is known as

gel permeation chromatography. At the present time no
semirigid gel is available which is truly compatible with
aqueous solvents, while free from adsorption. Semirigid
gels can be chemically and physically modified to render
them wettable. Mild sulfonation (17,18) (Aquapak®, Waters
Associates), chloromethylation + esterification, surfactant wetting, or mixed solvents (water-methanol) may be
used for an appropriate application, but none of these
treatments is universally applicable. The most widely
used semirigid gel is a highly cross-linked polystyrene
called Styragel®, which is provided by Waters Associates.

Rigid Gels. Although some of the substrates in this
category are glasses rather than gels, it is convenient
to remain consistent in the nomenclature and refer to
them as rigid gels. As a group, rigid gels offer fixed
pore sizes desirable for accurate physical measurement.
They also offer high column permeability and average capacity (i.e., 0.8-1.1). They can be both hydrophylic and
lypophylic. Most rigid gels offer V_S/V_M ratios of 0.8-
1.3 (although one commercial material falls below 0.5).
Plate efficiencies obtainable range between 100 and 500
ppf, as compared to 500-2000 ppf with semirigid gels. A
further disadvantage experienced with rigid gels is that
greater peak broadening occurs at higher flow rate than
is experienced with the semirigid material (19). Increase
in the resistance to mass transfer in stagnant pools of
moving phase obviously limits speed.

At the present time rigid substrates are either silica
gels or glasses, and as such exhibit adsorption. The
silicas contain hydroxyl groups which retard polar species
excessively, interfering with size separation (K_O becomes
greater than 1). Although this effect may be useful for
a chromatographic separation, it introduces error into
the elution volume versus size relationship. Suppressors,
such as triethylene glycol, may be added to the solvent
to reduce adsorption, but this treatment, in turn, com-
plicates the system. The preferred way is to deactivate
the silica chemically. The performance of chemically
deactivated silica has been reported by Bombaugh et al.
(20). Figure 7.7 shows chromatograms of polyvinyl

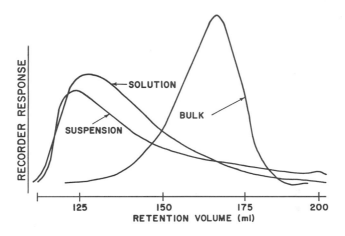

<u>Figure 7.7</u>. GPC chromatograms of polyvinyl alco-
hol from various methods of production (<u>20</u>).
<u>Conditions</u> - column: 4 ft x 3/8 in. of <u>deactivated</u>
Porasil®, 4 each in series packed with P-1000,
P-400, P-250; solvent: water at 65°C; sample:
0.25% for 120 sec - 5 mg; flow: 1 ml/min.

alcohol that eluted from the deactivated Porasil®. (This
material and various others, including polysodium silicate, which permanently adsorbs on untreated silicas,
eluted without evidence of adsorption from the deactivated Porasil®.)

Controlled-porosity glass, developed by Haller (<u>21</u>),
also exhibits adsorption, but the retardation is due to
a different mechanism from that operating with silica.
Consequently, conventional adsorption suppression techniques do not apply. Controlled-porosity glass offers a
unique feature, however, in that it presents extremely
uniform pore sizes as a result of its method of manufacture. This feature increases the selectivity term in
the resolution equation, making this packing a reasonable
candidate for applications in which high resolution or
narrow distribution is required. When controlled-porosity glass is applied to samples representing a wide

distribution of molecular weights, however, materials
with a wide range of porosities must be blended together,
which negates this advantage.

Commercially available packings in each class are listed in Table 7.3.

2. Selection of a Column Packing

Packing selection is based on the permeation range of
the gel, which is contained between the exclusion limit
and the total permeation limit indicated by the calibration curve. Material producing the top calibration curve
(C) in Figure 7.8 would not be suitable for resolving the
components represented, since all material permeates virtually all of the gel, offering poor resolution.   The
material producing curve A would also be a poor choice,

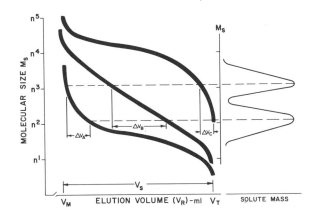

Figure 7.8.   Choosing a GPC column.

since a large portion of the distribution would be excluded from the gel. The material producing curve B,
however, would be ideal, since the solute pair falls
within the linear permeation range of the gel, giving
maximum $\Delta V_R/\Delta M_W$.

Although the permeation range is the property of interest, many suppliers list only the exclusion limit.
Exclusion limits, however, are not expressed in the same
units as are used in Table 7.3. For Styragel® the

Table 7.3

Commercially Available Column Packings for Exclusion Chromatography

Designation	Pore Diameter, Å	Molecular Weight Exclusion Limit	Distributor[a]
I. Soft Gels – for Low-Velocity LC			
A. Dextran gels			
Sephadex® G-10	. . . .	700[b]	(1)
Sephadex® G-15	. . . .	1,500[b]	(1)
Sephadex® G-25	. . . .	5,000[c]: 5,000[d]	(1)
Sephadex® G-50	. . . .	10,000[c]: 10,000[d]	(1)
Sephadex® G-75	. . . .	50,000[c]: 70,000[d]	(1)
Sephadex® G-100	. . . .	100,000[c]: 150,000[d]	(1)
Sephadex® G-150	. . . .	150,000[c]: 400,000[d]	(1)
Sephadex® G-200	. . . .	200,000[c]: 800,000[d]	(1)
Sephadex® LH-20	. . . .	e	(1)
B. Polyacrylamide gels			
Bio-Gel® P-2	. . . .	2,000[d]	(2)
Bio-Gel® P-4	. . . .	4,000[d]	(2)
Bio-Gel® P-6	. . . .	5,000[d]	(2)
Bio-Gel® P-10	. . . .	17,000[d]	(2)
Bio-Gel® P-30	. . . .	50,000[d]	(2)
Bio-Gel® P-60	. . . .	70,000[d]	(2)
Bio-Gel® P-100	. . . .	100,000[d]	(2)

255

Table 7.3 (cont.)

Designation	Pore Diameter, Å	Molecular Weight Exclusion Limit	Distributor[a]
Bio-Gel® P-150	. . .	150,000[d]	(2)
Bio-Gel® P-200	. . .	300,000[d]	(2)
Bio-Gel® P-300	. . .	400,000[d]	(2)
C. Agarose Gels			
Bio-Gel® A-0.5	. . .	500,000[d]	(2)
Bio-Gel® A-1.5	. . .	1,500,000[d]	(2)
Bio-Gel® A-5	. . .	5,000,000[d]	(2)
Bio-Gel® A-15	. . .	15,000,000[d]	(2)
Bio-Gel® A-50	. . .	50,000,000[d]	(2)
Bio-Gel® A-150	. . .	150,000,000[d]	(2)
Sepharose® B	. . .	3,000,000[c]	(1)
Sepharose® 2B	. . .	20,000,000[c]	(1)
Sag® (Ago-Gel)-10	. . .	250,000[d]	(3,4)
Sag® (Ago-Gel)-8	. . .	700,000[d]	(3,4)
Sag® (Ago-Gel)-6	. . .	2,000,000[d]	(3,4)
Sag® (Ago-Gel)-4	. . .	15,000,000[d]	(3,4)
Sag® (Ago-Gel)-2	. . .	150,000,000[d]	(3,4)
D. Polystyrene Gels			
Bio-Beads® S-X1	. . .	3,500[f]	(2)
Bio-Beads® S-X2	. . .	2,700[f]	(2)

Table 7.3 (cont.)

	Pore Diameter, Å	Molecular Weight Exclusion Limit	Distributor[a]
Bio-Beads® S-X3	...	2,100[f]	(2)
Bio-Beads® S-X4	...	1,700[f]	(2)
II. Semirigid Gels - for High-Velocity LC			
A. Polystyrene Gels			
Styragel® 39720	...	60 Å	(5)
Styragel® 39721	...	100 Å	(5)
Styragel® 39722	...	350 Å	(5)
Styragel® 39723	...	700 Å	(5)
Styragel® 39724	...	2,000 Å	(5)
Styragel® 39725	...	5,000 Å	(5)
Styragel® 39726	...	15,000 Å	(5)
Styragel® 39727	...	50,000 Å	(5)
Styragel® 39728	...	150,000 Å	(5)
Styragel® 39729	...	700,000 Å	(5)
Styragel® 39730	...	5,000,000 Å	(5)
Styragel® 39731	...	10,000,000 Å	(5)
Bio-Beads® S-X8	...	1,000[f]	(2)
Aquapak® A-440	...	100,000[f]	(5)
B. Polyvinylacetate gels			
Merck-o-gel-OR-750	...	750[f]	(6)

257

Table 7.3 (cont.)

Designation	Pore Diameter, Å	Molecular Weight Exclusion Limit	Distributor[a]
Merck-o-gel-OR-1500	...	1,500[f]	(6)
Merck-o-gel-OR-5000	...	5,000[f]	(6)
Merck-o-gel-OR-20,000	...	20,000[f]	(6)
Merck-o-gel-OR-100,000	...	100,000[f]	(6)
Merck-o-gel-OR-1,000,000	...	1,000,000[f]	(6)

III. Rigid Materials - for High-Speed LC

A. Porous silica

Porasil®-60	...	60,000[f]	(5)
Porasil®-250	...	250,000[f]	(5)
Porasil®-400	...	400,000[f]	(5)
Porasil®-1000	...	1,000,000[f]	(5)
Porasil®-1500	...	1,500,000[f]	(5)
Porasil®-2000	...	2,000,000[f]	(5)
Merck-o-gel® Si-150	150	50,000[f]	(6)
Merck-o-gel® Si-500	500	400,000[f]	(6)
Merck-o-gel® Si-1000	1000	1,000,000[f]	(6)

B. Porous glass

Bio-Glas® 200	200	...	(2)
Bio-Glas® 500	500	...	(2)

Table 7.3 (cont.)

Designation	Pore Diameter, Å	Molecular Weight Exclusion Limit	Distributor[a]
Bio-Glas® 1000	1000	· · ·	(2)
Bio-Glas® 1500	1500	· · ·	(2)
Bio-Glas® 2500	2500	· · ·	(2)
CPG 10-75	75	28,000[c]	(7,5)
CPG 10-125	125	48,000[c]	(7,5)
CPG 10-175	175	68,000[c]	(7,5)
CPG 10-240	240	95,000[c]: 120,000[f]	(7,5)
CPG 10-370	370	150,000[c]: 400,000[f]	(7,5)
CPG 10-700	700	300,000[c]: 1,200,000[f]	(7,5)
CPG 10-1250	1250	550,000[c]: 4,000,000[f]	(7,5)
CPG 10-2000	2000	1,200,000[c]: 12,000,000[f]	(7,5)

[a] Distributor:

(1) Pharmacia Fine Chemicals, Inc., Uppsala, Sweden; also Piscataway, N.J. 08854.

(2) Bio-Rad Laboratories, Richmond, Calif., 94802

(3) Seravac Laboratories (Pty), Ltd., Holysport Maiden Head, Berkshire, England.

(4) Mann Research Laboratories, Inc., New York, N.Y. 10006.

(5) Waters Associates, Inc., 61 Fountain St., Framingham, Mass. 01701.

(6) Merck AG, Darmstadt, West Germany.

(7) Corning Glass Works, Corning, N.Y. 14830.

Table 7.3 (cont.)

b Determined with polyethylene glycols (aqueous).

c Determined with soluble dextrans.

d Determined with peptides and/or proteins (aqueous).

e Exclusion limit depends on solvent employed and resultant degree of swelling.
Some solvents that can be used are water, methanol, ethanol, chloroform,
n-butanol, dioxane, tetrahydrofuran, N,N'-dimethylformamide, acetone, ethyl
acetate, and toluene.

f Determined with polystyrene.

g Extended chain lengths of polystyrene. Multiply by 41 to convert these
values to the approximate molecular weight of polystyrene.

exclusion limit is expressed as the extended chain length
of polystyrene in angstroms (Å). For most other materials
the exclusion limit is expressed as the molecular weight
of the standard used to calibrate the gel. Standards include a variety of materials, however, including polystyrene, polydextrans, polyglycols, polypeptides, and proteins, as shown in Table 7.3. When a portion of the
distribution is excluded from the gel, an anomalous distribution curve is produced. The materials which are
excluded elute simultaneously and produce a peak. This
is demonstrated in the chromatogram of crude oil in Figure 7.9. What appears as a peak is only that portion of
the distribution which was excluded from the gel and eluted as a narrow band. To resolve the excluded portion of

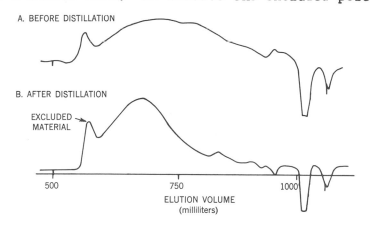

Figure 7.9. GPC characterization of crude oil (2).
Conditions - column: 20 ft x 3/8 in. o.d. of 500 Å
Poragel®.

the distribution would require an additional column with
a higher exclusion limit. As a matter of interest, it
may be a useful practice to select a gel so as to deliberately exclude a portion of the distribution and thereby
accentuate the excluded portion as a distinct peak. Peak
height, rather than the area under the tail, can then be
measured as an indication of the amount present. It is
simpler to measure peak height as an indication of the
area than to integrate the area under a long, flat tail.

3. Column Preparation

Column packing methods vary with the nature of the gel.
Semirigid gels, most widely used in high-performance gel
permeation chromatography, may be packed by the method
described by Moore (22). Uniform particles, between 37
and 75 μ in diameter, are suspended in several volumes of
a solvent mixture, such as perchloroethylene and toluene,
which are blended to match the density of the resin. The
constant-density slurry, free of particle segregation, is
then propelled by moving the solvent into the column as
a uniform plug and is compacted into a uniform bed. A
precolumn, approximately 1.5 times the length of the analytical column, is used to contain the slurry. The solvent is pumped at a flow rate sufficient to generate several hundred pounds of pressure drop. When constant
pressure is achieved, the solvent is circulated until the
gel is compacted into a uniformly dense bed. Optimum
pressure is determined empirically for a given batch of
gel.

Many practitioners of GPC elect to purchase ready-
packed columns rather than to make the investment in
equipment and technology required to pack their own columns. Ready-packed columns are commercially available
from Waters Associates, who produce Styragel® under a
license from Dow Chemical Company. A list of columns is
shown in Table 7.4.

The method used to pack soft, swellable gels differs
from that for rigid gels in that special effort must be
invested in maintaining uniform suspension of the slurry.
Packing must be carried out at lower transport velocity.
No differentiation need be made between lypophilic and
hydrophilic gels; these have a similar solvent reaction,
as long as they are suspended in a compatible solvent
which offers the same solvent-regain capacity as the solvent used in the separation. (CAUTION: A solvent used
to replace the one in which the column was packed must
not shrink the gel; otherwise, the soft gel will collapse,
producing voids in the column. Conversely, a change to
a solvent which markedly increases the swelling of the
gel can increase pressure drop or even plug the column.)

Table 7.4

Dimensions of Commercially Available
Ready-Packed Columns[a]

Nominal Permeation Ranges Available, Å[b]	Available Efficiencies in Each Permeation Range, ppf	Available Column Dimensions at Each Permeation Range and Each Efficiency
60	>450	3/8" x 4'
100-10	>700	3/8" x 3'
200-30		3/8" x 2(2')
500-40		1" x 4'
10^3-70		1" x 3'
$3(10)^3$-10^2		2.4" x 4'
10^4-$3(10)^2$		
$3(10)^4$-$5(10)^2$		
10^5-$2(10)^3$		
$3(10)^5$-$5(10)^3$		
10^6-10^4		

[a] Available from Waters Associates, Inc., Framingham, Mass.

[b] Å = extended chain length of polystyrene; 41 MW units/Å.

Numerous packing techniques have been reported. At the present time, Pharmacia, producer of Sephadex®, recommends an incremental addition which requires no column extension (23). The slurry of solvent-swollen gel is poured into the column, which is partially filled with solvent. The gel is allowed to settle until a gel bed forms of sufficient thickness to retard the flow of solvent from the column; the exit valve is then opened to permit the solvent to drain at a uniform rate. Slurry is then added to the column until a bed of the desired height is obtained.

Highest efficiencies obtained with soft gels were reported by Heitz (16), using 4-in.-diameter columns packed

with 2% cross-linked polystyrene. To obtain a uniform
bed, Heitz developed a highly sophisticated procedure in
which the column was rotated at random rates and for random duration in alternate directions. The procedure was
used to prepare the columns which made the separation
shown in Figure 7.5. Most practitioners, however, are
content to pack soft gels into columns by the method described by Pharmacia, using their own modifications. The
high capacity factor afforded by the soft gels markedly
overcomes the negative effect of random dispersion produced by a less efficient column, so long as the separations are made at low flow rates.

In contrast to the methods for packing soft and semirigid gels, rigid gels may be dry-packed by the conventional procedure widely used in gas chromatography and
described in Chapter 5 as a preferred method of packing
columns for adsorption and partition chromatography. Although many variations have been tried, columns with no
more than 100-500 ppf have been obtained by this approach.

4. Equipment

The discussion of equipment in this chapter is limited
to the aspects which are unique to GPC.

Column Diameter. The diameters of columns used in GPC
have been larger than those commonly employed in partition or adsorption chromatography. The most popular size
currently is 0.303 in. i.d. Early workers found that, as
column diameter was reduced, efficiencies dropped significantly, and, surprisingly, as diameters were increased to
greater than 2 in., efficiencies improved. Preparative
columns 2 1/2 in. i.d. of Styragel® have been packed routinely with more than 2000 ppf. Working with soft gels,
Heitz reported that it was very difficult to obtain high
efficiencies with columns 10 mm in diameter; however, he
routinely obtained efficiencies greater than 2000 ppf
with columns 60 mm in diameter (16).

It follows that the large column diameters and greater
column lengths required for high resolution by GPC impose
specific demands on the design of the chromatographic system, in that the solvent system must accommodate the

volume demands of the column. Either a continuous pump must be used, or, if a syringe-type pump is employed, it must accommodate at least two to three times the displacement volume of the column to avoid a pulse surge during a run.

Sample Injection. Gel permeation chromatography equipment should have both a valve loop and a septum injector for general use. The valve loop is necessary to accommodate viscous samples. Most polymeric samples must be injected as 0.25-1% solutions. At higher concentrations, high solution viscosity causes increased pressure drop and unnecessary band spreading due to viscous streaming on the trailing side of the solute band. Typical column systems for molecular weight determination in GPC are 16 ft x 3/8 in. Therefore, sample volumes range between 0.25 and 2 cc. Because of the large sample aliquots and the high solution viscosity, such samples cannot be injected into a high-pressure system with a typical microliter (liquid) syringe. However, for conventional analytical work with small molecules (i.e., those which are not viscous in concentrated solutions) samples may be introduced through a typical septum injector, using a microliter syringe. The valve loop injector is also a convenient method of introducing large sample loads for preparative work.

5. Operating Conditions

Choosing a Solvent. A GPC solvent should dissolve the sample and be sufficiently similar to the gel so as to wet the gel and prevent adsorption. When working with a soft gel, the solvent must also swell the gel, since the pore size of soft gels is a function of the solvent imbibed. In the ideal case, the moving solvent, the trapped solvent, and the gel should interact identically with the solute molecule to ensure that its movement into the pore will occur strictly by diffusion. In practice, however, this rule is not followed rigidly, as indicated by the data in Table 7.5, showing the solvents most commonly used in gel permeation and gel filtration chromatography. For example, N,N'-dimethylformamide

Table 7.5

Most Commonly Used GPC Solvents

I. Physical Properties

Solvent	Boiling Point, °C	Density	Viscosity	Refractive Index	Use Temperature, °C	Use
Tetrahydrofuran	66	0.8892	0.51 at 25°	1.4070[20]	RT-45	General polymer and small molecules
1,2,4-Trichloro-benzene	213	1.4634[25]	0.50 at 135°	1.5524[25]	130-160	Polyolefins
Toluene	110.6	0.866[20]	0.52 at 25°	1.4893[24]	RT-70	Rubbers and elastomers
m-Cresol	202	1.034[20]	16.9 at 20°	1.5348[20]	30-135	Polyesters and polyamides
N,N'-Dimethyl-formamide	153	0.9445[25]	0.90 at 25°	1.42803[25]	RT-85	Polyurethanes, acrylates, cellulose esters, acrylonitrile
Chloroform	61.2	1.489[20]	. . .	1.4476[20]	RT	Epoxies, silicones, vinyl polymers, small molecules
1,1,2,2-Tetra-chloroethane	146.5	1.58658[25]	. . .	1.49419[20]	RT-100	Low-molecular-weight compounds
Trifluoroethanol	73.6	1.38232[5]	0.9 at 38°	1.2907[20]	RT-40	Some polyamides
Water[a]	100	0.9999[20]	1.0 at 20°	1.3330[20]	RT-65	Polyelectrolytes and biological materials

[a] Aqueous buffers may be required to stabilize pH, molecular composition, or molecular size.

Table 7.5 (cont.)

Most Commonly Used GPC Solvents

II. Safety Information

Solvent	Flash Point, °C	Acute LD50, mg/kg	Acute Inhalation L (ct) 50, ppm
Tetrahydrofuran	33.8	Dangerous	200
1,2,4-Trichloroben- zene	110	Dangerous	75
o-Dichlorobenzene	79	500	75
m-Cresol	86	Dangerous	5
Toluene	8	Dangerous	6000
N,N'-Dimethyl- formamide	67.4	Dangerous	100
Trifluoroethanol	40.5	240	4600
Carbon tetrachloride		Dangerous	25

does not meet the requirement described above in that it is not similar to the gel; nevertheless, it is frequently used with satisfactory results with samples which it alone dissolves.

Solvent Viscosity. Solvent viscosity is an important consideration in GPC, since high viscosity restricts diffusion and impairs resolution. The consideration is even more important with macromolecules, which have relatively low diffusion coefficients. The solvent must be compatible with the detector and must permit discrimination of the solute from the solvent. At the present time the most commonly used detector for GPC is the differential refractometer. Therefore, such materials as toluene, trichlorobenzene, and m-cresol are widely used as solvents; with this technique, they are the preferred solvents for many organic soluble polymers and permit ready discrimination with the RI monitor. They find little application in gel filtration chromatography, however, not only because of their solvent properties, but also because their strong UV adsorption renders them unsuitable for use with the UV detector commonly employed.

The solvent must be compatible with the system hardware. Halogen salts, which are commonly added to biological materials to serve as stabilizers and to control the pH of aqueous systems, can cause pit corrosion in stainless steel components. Therefore, sulfate and phosphate buffers should be substituted whenever possible. The role of salt as an electrolyte is important in exclusion chromatography since some molecules change size when the solvent composition or the electrolyte concentration is changed. In addition, changes in electrolyte strength can alter the pore size of soft gels.

The effect of solvent composition on elution behavior is illustrated by the family of calibration curves shown in Figure 7.10. The pore size of the rigid gel, Porasil®, is invariant to solvent; nevertheless, the calibration curves of the dextrans shift with electrolyte concentration in the solvent, showing that the dextran molecule is more extended in water than in the salt solution. Details of curve shift are described in the caption. The apparent molecular weight of polyelectrolyte has been known to change by a factor of 10 between water and the 0.1 $\underline{N}$ electrolyte solution. This anomaly must be considered in deriving size information from the chromatogram.

Effect of Temperature. The primary need for elevated temperature in GPC is to dissolve materials not soluble at room temperature, including polyolefins, polyesters, and some polyamides not soluble in trifluoroethanol. Virtually all other materials may be separated at ambient temperature. However, increasing the temperature above ambient reduces solution viscosity and increases molecular diffusion, thereby increasing either resolution or speed.

C. HIGH RESOLUTION OF DISCRETE SPECIES BY GEL PERMEATION CHROMATOGRAPHY

1. Utility of High-Resolution Gel Permeation Chromatography

In the fractionation of macromolecules, GPC has enjoyed

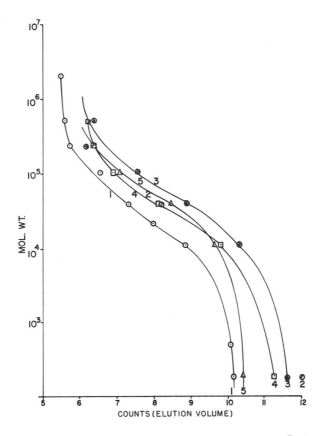

<u>Figure 7.10</u>. Calibration curves, Porasil®-400 (C).
<u>Conditions</u> - 1: In water, dextrans are extended
and elute at a lower V_R (i.e., they are larger, relative to their molecular weight). 2-3: Salts compress molecule, but do not suppress adsorption.
Molecules appear smaller than they actually are.
4-5: Salts compress molecules; glycols suppress
adsorption. Both effects are suppressed, so that
lines 4 and 5 relate most accurately to the size
of the dextran molecules.

wide usage. During the past few years it has been applied
increasingly by the polymer industry to solve a variety

of problems relating to molecular size or molecular
weight distribution. These studies have dealt with polymer synthesis (24), polymer blending (25), and polymer
degradation (26).

The technique can be equally useful, however, to those
working with low-molecular-weight substances, such as
oligomers, monomers, and many nonpolymeric substances.
In fact, it is in the low-molecular-weight region that
highest resolution can be obtained by GPC. This is evident in the triglyceride separation shown in Figure 7.11,
where the materials differing in molecular weight by an
equal amount ($\Delta M_W = 40$) experienced highest resolution at
the low-molecular-weight (high k') end of the distribution. There is no lower limit to the size range of the

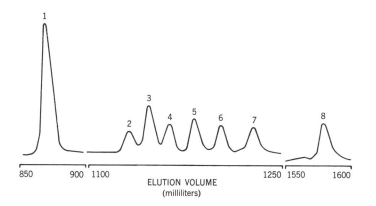

Figure 7.11. High-resolution GPC separation of
triglycerides (9,10). Conditions - column: 160
ft x 3/8 in. o.d. of 500 Å gel; flow rate: 0.4
ml/min; peak identification: 1, polystyrene; 2,
triarachidin; 3, tristearin; 4, tripalmitin; 5,
trimyristin; 6, trilaurin; 7, tricaprin; 8, o-
dichlorobenzene.

molecules that GPC can separate. Hendrickson has reported separations of light gases by GPC (11). Using only a
12-ft column, he separated 12 components, covering molecular weight ranges between 76 and 500.

As solute molecular weight is increased, higher resolution is needed to separate components differing by an
equal amount. This is clearly illustrated by the hydrocarbon separation shown in Figure 7.12. Base-line resolution is achieved between materials differing by one

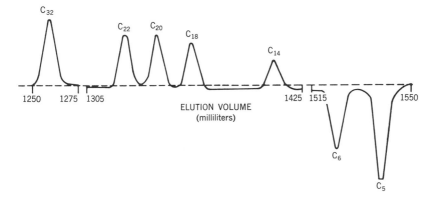

Figure 7.12. High-resolution GPC separation of
hydrocarbons (9,10). Conditions - column: 160
ft x 3/8 in. of 500 Å gel; solvent: tetrahydrofuran; sample: normal hydrocarbons, 3 mg each;
flow: 0.4 ml/min.

carbon number in the C_5-C_{10} region. A difference of two
carbon numbers is required for base-line separation in
the C_{16}-C_{24} region, and a four-carbon-number difference
at C_{32}. Since peak widths are equal, one can readily
predict the peak capacity and resolution capability of
the system (8).
   The calibration curve in Figure 7.13 can be informative. First, it shows that the molecular weights of
hydrocarbons and triglycerides relate linearly to V_R.
Since peak widths with this system are 14 ml, one can
measure the peak capacity of the system and predict quite
reliably where a component of a given molecular weight
will elute. Conversely, a peak eluting at known elution
volume can be assigned a molecular weight with reasonable
confidence. If a subsequent separation by adsorption

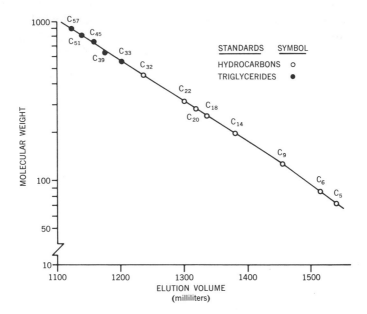

Figure 7.13. High-resolution GPC system calibration curve, showing the molecular weights of hydrocarbons and triglycerides relative to V_R as a linear log function (9,10).

yields a multiple-component fraction, the practitioner has the advantage of knowing that all the components of the fraction eluted at the same elution volume and are, therefore, of similar molecular weight.

2. High Resolution Using Recycle

The high-resolution separations shown in Figures 7.11 and 7.12 were made on a comparatively long Styragel® column, which represented a relatively large capital investment. An alternative to long columns as a means of increasing resolution in GPC is to recycle the solute through the column (9,27). This approach is particularly reasonable in GPC, since K_O values do not exceed 1. As a result, the maximum elution volume is fixed by the capacity of the system. Late peaks common to sorption

chromatography need not be considered. The C_{14}-C_{16} hydrocarbon separation shown in Figure 7.14 was made using 20
cycles at 3 ml/min on a 16-ft column. In comparison to

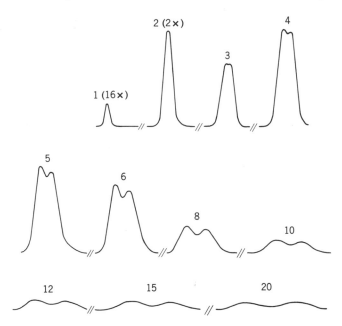

Figure 7.14. Recycle separation of hydrocarbons
(C_{16} and C_{14}) (9). Conditions - column: 16 ft x
3/8 in. of 500 Å Poragel®; solvent: toluene; sample size: 10 μl. Numbers over peaks represent
cycles.

the 160-ft system, this column was one-tenth the length,
was run at 7.5 times the flow rate, and made the separation in one-tenth the time by using recycle. The net
gain was achieved because with recycle it was possible to
increase effective column length without increasing pressure drop.

Recycle chromatography, first considered in gas chromatography (28), was introduced into gel filtration by
Porath and Bennich (27). Bombaugh et al. (9) demonstrated that a small-volume reciprocating pump used in commercially available GPC equipment was capable of recycle

operation. These workers showed that, by connecting the outlet of the detector to the suction of the pump with a small-diameter tubing, separations like the one shown in Figure 7.15 were possible. Data from an analysis of the chromatogram are given in Table 7.6.

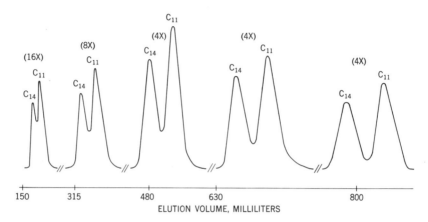

Figure 7.15. Recycle separation of hydrocarbons (C_{11} and C_{14}) (9). Conditions - column: 16 ft x 3/8 in. of 500 Å Poragel®; solvent: toluene; flow: 1.5 ml/min; sample size: 10 µl.

Equipment for recycle operation differs from conventional liquid chromatography equipment in that it imposes a concern for band spreading in the entire chromatographic system. This involves not only the detector and transport tubing, but also the pump (29,27). Syringe or large piston pumps (often referred to as pulseless pumps) are unsuitable for recycle operations, since they would spread the peak over the volume of displacement. For a recycle system to be practical, the extra-column band spreading must be small relative to the band spreading of the column. With a 0.303 in.-diameter column, commonly used in GPC, and with lengths ranging between 16 and 40 ft, these criteria are easily met.

An additional consideration arises because a recycle system is a closed system with a finite volume; hence, the fast-moving material will eventually overtake the slower-moving material and remix. To prevent peak

Table 7.6

Separation Data from Recycle Operation

Cycle	V_R	w	N	ppf	V_R	w	N	ppf	Relative Retention α	σ
1	157.5	7.6	6,850	428	161.6	7.0	8,460	529	1026	2.2
2	319.3	10.9	13,750	430	327.4	11.1	13,820	432	1025	2.9
3	480.6	13.8	19,510	406	494.6	14.4	18,890	395	1029	4.0
4	643.5	15.8	26,460	413	661.0	16.6	25,220	394	1027	4.3
5	805.6	18.6	29,990	375	827.4	18.8	31,160	389	1027	4.7

Conditions:

Sample: 10 μl of C_{11} and C_{14} hydrocarbons

Columns: 16 ft of 500 Å gel

Solvent: Toluene

Flow rate: 1.5 ml/min

overlap, a means must be provided to permit the operator
to remove a portion of the distribution before overlap
can occur (29). This is illustrated by the chromatograms
in Figure 7.16, where Triton® X-45 is passed through six
cycles. Since the low-molecular-weight end is resolved
first, it is preferable to draw off the resolved components and continue to recycle the rest until resolution

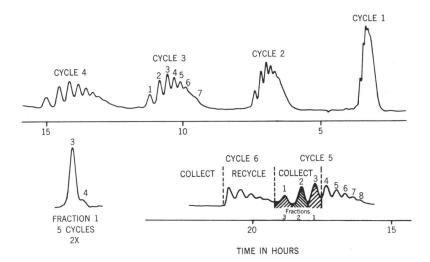

Figure 7.16. Effect of cycle number on resolution
(10). Conditions - column: 15 ft of 60 Å Poragel®;
solvent: tetrahydrofuran; sample: 30 μl Triton®
X-45, concentration 50%.

is complete. To accomplish this removal accurately, it
is obviously necessary to recycle the solute through the
detector.

Peak width increases with column length and can, therefore, be expected to increase with cycle number. Peak
width on a particular cycle is described by the equation

$$w_{\gamma_n} = w_0(\gamma_n)^{1/2}$$

where γ_n = cycle number. Since distance between peaks
increases as a function of the number of cycles (γ_n), the
width of an unresolved mixture on a particular cycle

becomes

$$w_{\gamma n} = (w_0^2\gamma_n + w_0^2\gamma_n^2)^{1/2}$$

The maximum number of cycles is then determined by the initial capacity of the system and the breadth of the solute distribution (29, 30). To prevent overlap, either the side of the distribution must be removed, as stated earlier, or the column length must be increased. It is usually preferable to remove the portion from the high k' side of the distribution, where resolution is greatest. (In GPC, low-molecular-weight materials show the highest k'.)

As just mentioned, peak width increases with column length and with the cycle number. As peak width increases, the concentration of the solute in the solvent is lowered, resulting in reduced sensitivity at the detector. It may be necessary, therefore, to increase the sample load to compensate for the reduced sensitivity which accompanies the increase in the effective length of the column by re-cycle.

Effect of Recycle on Resolution at Heavy Load. Recy-cle affords a particular advantage for preparative chro-matography. The effect of sample load on resolution is shown in Figure 7.17. Figure 7.18 shows that a 5-gram load on a 2 1/2-in.-i.d. column produces a badly distorted peak which may be misinterpreted as a trimodal distribu-tion. After only three cycles, the sample is resolved as a true bimodal mixture. Column length is programmed as needed to obtain the resolution desired. More complete data showing the effect of load on resolution at up to three cycles are shown in Figure 7.19. The deleterious effect of overload on resolution is evident. However, by judicious use of overload plus recycle, throughput can be maximized to considerable advantage.

Effect of Flow Rate on Resolution with Recycle. A similar affect on resolution applies to flow rate. The data plotted in Figure 7.20 show the relationship between resolution and flow rate at each of three cycles. Reso-lution at 120 ml/min after three cycles is comparable to that at 14 ml/min after one cycle. By increasing the

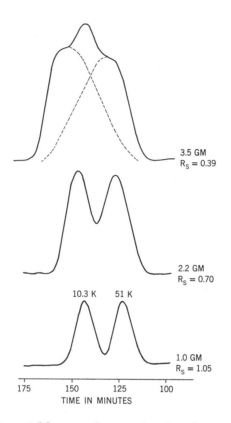

<u>Figure 7.17.</u> Effect of sample load on resolution
(<u>10</u>). <u>Conditions</u> - column: 4 ft of 2.5 x 10⁴ Å
Styragel®; sample: polystyrene mixture
51K + 10.3K (1-1); concentration: 10 mg/ml; injection: varied.

flow rate and using recycle, throughput at equal resolution is tripled.

Increased flow rate is desirable for increased throughput, when system capacity provides adequate resolution
before overlap occurs and when the system is attended to
make the necessary fraction removal. Conversely, it is
often advantageous to decrease the flow rate to provide
increased resolution per cycle, and to let the system
recycle overnight unattended, if it can run through four

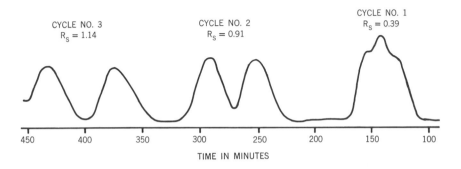

Figure 7.18. Effect of recycle on resolution at
heavy load (10). Conditions - column: 4 ft of
2.5 x 10⁴ Å Styragel®; solvent: toluene; sample:
polystyrene mixture 51K + 10.3K(1-1); concentration: 10 mg/ml; injection: 350 ml; load: 3.5 g.

or five cycles without overlap. In this way, maximum
resolution per column per cycle is obtained. Although
this procedure represents an increase in elapsed time,
it permits the equipment to work overnight unattended
and increases throughput per working man-hour.

A schematic drawing of a recycle system is shown in
Figure 7.21. This system permits the recycle and fraction removal steps that have been described. Facilities
are also provided to flush both the inlet "T" and the
connecting tubing. This sampling section may contain
traces of solute which would be reinjected by the fresh
solvent entering during the "draw-off" cycle and would
appear as a spurious peak.

D. INTERPRETATION OF GEL PERMEATION CHROMATOGRAMS

Gel permeation chromatography has been used primarily
as a means for analyzing the molecular weight distributions of polymeric materials, whereas gel filtration
chromatography has served chiefly as a preparative separation technique. Either procedure is suitable for both
fractionation and size analysis. For molecular size

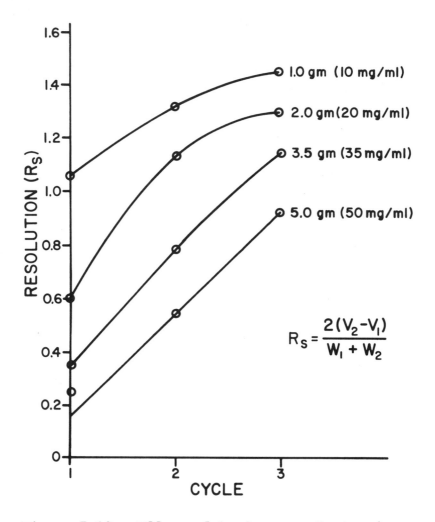

Figure 7.19. Effect of load on resolution (constant volume) (28). Conditions - column: 4 ft of
2.5 x 10⁴ Å Styragel®; solvent: toluene; sample:
polystyrene mixture 51K + 10.3K (1-1); injection:
100 ml; flow: 14.4 ml/min.

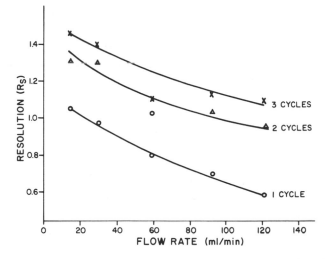

Figure 7.20. Effect of flow rate on resolution
(28). Conditions - column: 2.5 x 10^4 Å Styragel®;
solvent: toluene; sample: polystyrene mixture
51K + 10.3K (1-1); injection: 100 ml; load: 1 g.

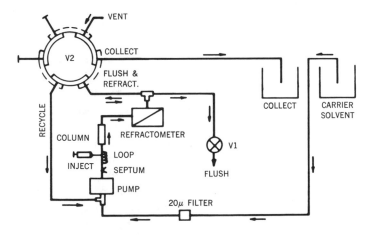

Figure 7.21. Schematic diagram of recycle operation (10).

analysis, a relationship must be established between the chromatograms and the molecular size or, more properly, the molecular weight. This relationship can be established by means of a calibration curve. When resolution is adequate to obtain individual peaks for single species, the relationship may be reasonably straightforward, as illustrated in Figure 7.13. Deviations are known to occur when molar volume or size in solution does not correlate with molecular weight.

When working with polymeric systems, resolution to single species is neither practical nor necessary. Useful systems have been developed to relate the unresolved GPC envelope to an expression of molecular weight distribution by use of a calibration curve. Direct calibration of the GPC columns with narrow standards of known molecular weight is the preferred approach. When such standards are not available, it has been customary to relate the molecular weight of the solute to the chain length of the extended polymeric chain by the so-called Q factor. This technique (31), though subject to anomalies, has been in wide use, since it is simple to apply and is a practical expedient until valid calibration standards can be obtained.

Interpretation of the chromatogram requires a valid relationship between retention volume (V_R) and a stable molecular dimension. Size in solution is neither a stable nor an intrinsic dimension, since molecules change size with solute concentration and solvent composition. Therefore, when it is not possible to calibrate with valid standards of known molecular weight, the relationship between V_R and average molecular weight is at best only an approximation. To overcome the unpredictability of molecular size in solution, it is necessary to include an additional parameter, which, when combined with molecular weight, provides an intrinsic dimension related to the molecular radius of gyration. The dimension proposed by Benoit (33) is hydrodynamic volume, which is the product of molecular weight times intrinsic viscosity. Intrinsic viscosity is the solution viscosity extrapolated to infinite dilution. At infinite dilution the molecule

matches the condition of the solute in a properly operated gel column. The viscosity provides a means of differ-
entiating between two molecules of the same molecular
weight but different spacial configurations.

Both Benoit (33) and Devries (34) demonstrated that a
single calibration line could be applied to polymers with
a variety of compositions and structural configurations
when hydrodynamic volume was plotted against GPC retention volume. Cazes (31,4) showed that for small, nonpolymeric molecules molar volumes correlate well with
retention volumes. Molar volumes can be calculated from
experimentally measured densities or estimated at the temperature of interest by means of critical properties.
Hendrickson and Moore (5) proposed that retention volume
could be related to effective carbon number, an empirically derived relationship based on the length of the
carbon-to-carbon bond. The retention volumes of 130 compounds with a variety of compositions containing a variety
of functional groups were measured to relate other functional bond distances to the carbon-carbon bond length.
By this approach, the contributions of all atomic group-
ings can be added together to obtain an effective chain
length from which retention volume can be predicted. The
relationship was used effectively by Spell in conjunction
with infrared spectometry to make composition assignments
directly from GPC chromatograms (7).

Although highly accurate curve interpretation requires
correction for geometric configuration, much meaningful
work can be done by direct calibration or by calibration
with secondary standards. For most accurate interpretation allowance must be made for curve anomalies which
frequently appear as skewness. Such distortions can be
caused by column overload, adsorptive effects, or a poorly designed system. When such anomalies are observed,
they should be identified and corrected.

References

1. J. C. Moore, J. Polymer Sci., Part A-2, 835 (1964).
2. K. J. Bombaugh and R. F. Levangie, Anal. Chem., 41,
    1357 (1969).

3. K. J. Bombaugh, W. A. Dark, and R. F. Levangie, Z. Anal. Chem., 236, 443 (1968).
4. J. Cazes and D. R. Gaskill, Separation Sci., 2, 421 (1967).
5. J. G. Hendrickson and J. C. Moore, J. Polymer Sci., Part A-4, 1967 (1966).
6. R. D. Law, Proceedings 4th Intern. GPC Seminar, Miami, May, 1967, p. 180, published by Waters Associates, Inc., Framingham, Mass.
7. J. L. Spell, Proceedings 4th Intern. GPC Seminar, Miami, May, 1967, p. 212, published by Waters Associates, Inc., Framingham, Mass.
8. K. J. Bombaugh, W. A. Dark, and R. F. Levangie, Separation Sci., 3 (4), 375 (1968).
9. K. J. Bombaugh, W. A. Dark, and R. F. Levangie, J. Chromatog. Sci., 7, 42 (1969).
10. K. J. Bombaugh and R. F. Levangie, Separation Sci., 5 (6), 1970.
11. J. G. Hendrickson, Anal. Chem., 40, 49 (1968).
12. L. R. Snyder and D. L. Saunders, J. Chromatog. Sci., 7, 195 (1969).
13. L. R. Snyder, J. Chromatog. Sci., 7, 352 (1969).
14. J. L. Waters, J. N. Little, and D. F. Horgan, J. Chromatog. Sci., 7, 293 (1969).
15. K. J. Bombaugh, Can. R/D, 2 (5), 41 (1969).
16. W. Heitz, B. Bomer, and H. Ullner, Makromol. Chem., 121, 102-116 (1969).
17. K. J. Bombaugh, W. A. Dark, and R. F. Levangie, "Applications of GPC to Aqueous Systems," in Proceedings 4th Intern. GPC Seminar, Miami, May, 1967, p. 33, published by Waters Associates, Inc., Framingham, Mass.
18. K. J. Bombaugh, W. A. Dark, and R. N. King, J. Polymer Sci., Part C, 21, 131 (1968).
19. R. N. Kelly and F. W. Billmeyer, Jr., Preprints Div. Petroleum Chem., Am. Chem. Soc., 15, No. 2, 157 (1970); Separation Sci., in press.
20. K. J. Bombaugh, W. A. Dark, and J. N. Little, Anal. Chem., 41 (10), 1337 (1969).
21. W. Haller, Nature, 207, 693 (1965).

22. J. C. Moore, U.S. Patent 3,326,875.
23. "Sephadex® in Gel Filtration," Pharmacia Fine Chemicals, Uppsala, Sweden.
24. W. A. Pavelich and R. N. Livigni, J. Polymer Sci., Part C, 215 (1967).
25. D. J. Harmon, J. Appl. Polymer Sci., 11, 1333 (1967).
26. J. G. Hendrickson, J. Appl. Polymer Sci., 11, 1419 (1967).
27. J. Porath and H. Bennich, Arch. Biochem. Biophys., 152 (1962).
28. A. J. P. Martin, in "Gas Chromatography," V. J. Coates, H. J. Noebels, and I. S. Fageson, eds., p. 237, Academic, New York, 1958.
29. K. J. Bombaugh and R. F. Levangie, J. Chromatog. Sci., 8 (10), 560-66 (1970).
30. K. J. Bombaugh, J. Chromatog., 53 (1), 27-35 (1970).
31. J. Cazes, J. Chem. Educ., 43, A567, A625 (1966).
32 Z. Grubisic, P. Rempp, and H. Benoit, J. Polymer Sci., Part B-5, 753 (1967).
33. Z. Gribisic-Gallot and H. Benoit, Proceedings 7th Intern. GPC Seminar, p. 65, Monte Carlo, 1969, published by Waters Associates, Inc., Framingham, Mass.
34. M. LePage, R. Beau, and A. J. Devries, J. Polymer Sci., Part C, No. 21, 119-130 (1968).

CHAPTER 8

Practice of Ion-Exchange Chromatography

Charles D. Scott

A. INTRODUCTION

Ion-exchange materials represent an important class of stationary phases used in liquid chromatography. When such materials are employed in chromatography, a reversible exchange of ions takes place between the stationary ion-exchange phase and the external liquid mobile phase. A difference in the affinity of the solute ions for the ion-exchange stationary phase allows a chromatographic separation.

Ion-exchange chromatography was one of the first chromatographic techniques used. Around the beginning of the twentieth century, ion-exchange materials began to gain prominence as media for use in water softening and treatment. At first, various natural minerals such as the zeolites were used; later, synthetic materials were prepared and tested.

The synthetic organic ion-exchange resins had their beginning in the 1930s (1). Numerous different synthetic resins were subsequently developed, and the extremely useful cross-linked polystyrene resins became available in the early 1940s. These resins, which were considered to be the standard materials for water treatment after World War II, are now the primary media used in ion-exchange chromatography.

The development of ion-exchange methods for separating the rare earths and various fission products (2,3) was extremely important in the field of atomic energy, and the extension of these methods to problems in biochemistry has made possible the solution of many extremely complex problems relating to the structure of proteins and nucleic acids (4,5). Preparative ion-exchange chromatography has also been used to isolate drugs, biochemicals, synthetic transuranium compounds, etc., on a production scale.

B.   ION-EXCHANGE MATERIALS

Although some liquids with ion-exchange properties can be successfully used as the stationary phase in liquid-liquid chromatography, the most useful ion-exchange materials are generally solids. In this discussion, only the solid exchangers are considered.

1.   Desirable Properties

An ideal ion-exchange material for chromatography should provide a stationary phase that is insoluble and chemically stable. The ion exchanger should have structural stability, and the particles should preferably be uniform spheres so that they will have good flow properties when they are packed in a column. The exchange capacity should be high, and ion-exchange sites should be monofunctional in nature.

2.   Types of Ion-Exchange Materials

There are many types of ion-exchange materials. These include naturally occurring minerals such as clays and zeolites, as well as several inorganic oxides; however, the solid ion exchangers most widely used in liquid chromatography are synthetic materials such as synthetic polymers and derivatives of cellulose.

3.   Ion-Exchange Resins

The synthetic resins constitute by far the most important class of ion-exchange materials for application in chromatography. These materials contain polar groups, acidic or basic in nature, which are introduced either before or after the polymerization stage.

Each ion-exchange resin consists of an insoluble, polymeric matrix that is permeable. This matrix or lattice contains fixed-charge groups and mobile counterions of opposite charge. These counterions can be exchanged for other ions in the external liquid phase. Thus ion-exchange chromatography is limited to the separation of substances that are at least partially in ionized form; however, physical sorption also occurs on the surface of the resin to some extent.

The ion-exchange sites are attached to the polymer chains and are thus distributed throughout the resin particle. The charged groups cannot have translational motion, and they are balanced by an equivalent number of mobile counterions. The type and the strength of the exchanger are determined by these groups.

Ion-exchange resins are called either cation-exchange resins or anion-exchange resins, depending on whether they have an affinity for cations or anions. Furthermore, each of these categories can be subdivided into strongly acidic and weakly acidic cation-exchange resins and strongly basic or weakly basic anion-exchange resins. Different active groups are present in various ion exchangers to give them the above-mentioned properties. The most important strongly acid cation-exchange resins contain sulfonic acid groups ($-SO_3^-H^+$) at the active sites (Table 8.1). Weakly acidic cation-exchange resins may contain active groups such as carboxylic groups ($-COO^-H^+$) in the internal structure. The most popular anion-exchange resin is the strongly basic polystyrene copolymer with quaternary ammonium exchange groups [$-CH_2N^+(CH_3)_3Cl^-$]. Weakly basic anion exchangers have functional groups such as $-N^+H(R_2)Cl^-$.

Hundreds of different types of ion-exchange resins are commercially available. (The selected bibliography at the end of this chapter covers the field in great detail.) However, we will consider only a few types that have been especially useful in liquid chromatography. Typical of these are the cross-linked polystyrene resins, which are actually copolymers of styrene and divinylbenzene (i.e., the divinylbenzene is cross-linked with the polymerized styrene to form an insoluble matrix). Available in either the anionic or the cationic form, these resins are chemically very stable and are essentially ideal ion exchangers in every respect. Other synthetic resins include those with an acrylic polymer lattice, a phenolic lattice, etc.

Table 8.1

Some Ion-Exchange Resins Used in Chromatography

Type	Name	Typical Functional Group
Strongly acidic cation exchanger	Dowex® 50 Amberlite® IR-120 Duolite® C-20 Aminex® A-6	$-SO_3^-H^+$
Weakly acidic cation exchanger	Amberlite® IRC-50 Duolite® CC-3	$-COO^-Na^+$
Strongly basic anion exchanger	Dowex® 1 Amberlite® IRA-400 Duolite® A-101 Aminex® A-27	$-CH_2N^+(CH_3)_3Cl^-$
Weakly basic anion exchanger	Dowex® 3 Amberlite® IR-45 Duolite® A-2	$-N^+H(R)_2Cl^-$

Trade Name	Manufacturer
Dowex®	Dow Chemical Co.
Amberlite®	Rohm and Haas Co.
Duolite®	Diamond-Shamrock
Aminex®	Bio-Rad Laboratories

4.   Nonrigid Ion-Exchange Materials

In order to overcome the problem of mass transport of macromolecules such as proteins in the ion-exchange matrix, another class of ion-exchange materials with a more porous matrix has been introduced. These exchangers have been prepared by introducing ionizable groups into polyacrylamide gels and carbohydrate polymers such as cellulose or dextran. The materials have a very low density of exchange sites and a hydrophilic type of matrix. They

tend to swell extensively in water, and the resulting nonrigid structure easily deforms in a flow field. In general, these materials are used to separate large molecules
of a biological origin, and the low density of exchange
sites permits exchange of such biochemicals under mild
conditions.

Somewhat different types of exchange groups are employed with these hydrophilic ion exchangers. Guanido-
ethyl (GE) is used for the strongly basic anion exchanger;
diethylaminoethyl (DEAE), for the weakly basic anion exchanger. Sulfoethyl (SE) and sulfomethyl (SM) are used
for the strongly acidic cation exchanger; carboxymethyl
(CM), as the weakly acid cation exchanger.

5. Structural Differences

Ion-exchange resins have structural differences (Figure 8.1) as well as chemical and size differences. Most
of the organic ion-exchange resins contain cross-linked
structural networks characterized by relatively small
openings (i.e., near-molecular size). These microreticular resins (Figure 8.1a) have a consistency similar to
that of a gel. Macroreticular ion-exchange resins (Figure 8.1b), on the other hand, contain not only the micro-
pores but also pores that may be several hundred angstroms
wide. Such resins have high internal surface areas and
relatively high porosities; the larger pores provide,
throughout the rigid pore structure, channels that render
the centers of these resin particles accessible to relatively large molecules and reduce the effects of solid-
phase mass transport.

Another structural type of ion-exchange resin is pellicular resin (Figure 8.1c), in which a solid core is
surrounded by a thin film of cross-linked material. This
type of resin has been shown to be extremely useful in
reducing solid-phase mass transport resistances and thus
accelerating the sorption process. However, the capacity
of the resin is extremely low, and chromatographic systems using this material are limited to very small amounts
of the individual solutes.

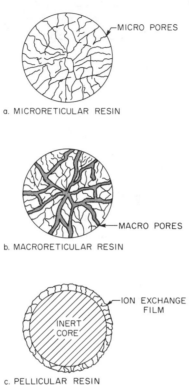

a. MICRORETICULAR RESIN

b. MACRORETICULAR RESIN

c. PELLICULAR RESIN

Figure 8.1. Structural types of ion-exchange resins.

6. Cross-Linkage

The degree of cross-linkage of a polystyrene-type resin is expressed as the proportion (in weight per cent) of divinylbenzene (DVB) present in the reaction mixture. With ion-exchange resins manufactured by the Dow Chemical Company, the degree of cross-linkage for a particular resin is given by the number that appears in the "-X(No.)" designation immediately following the trade name of the resin. For example, Dowex 1-X8 has a cross-linkage of 8%. An ion-exchange resin with less than 4% cross-linkage tends to be structurally unstable in an ion-exchange

column because of its tendency to collapse from the shear
force of flowing liquid; however, the effective pore size,
the permeability, and the tendency of such a resin to
swell in solution are quite high. On the other hand,
resin with a cross-linkage greater than 12% is characterized by satisfactory structural strength but has a much
smaller effective pore size, permeability, and tendency
to swell. Mass transport within such a resin is extremely slow, or even impossible, with larger molecules. The
resins of greatest utility have an intermediate amount of
cross-linkage (i.e., 4-12%), and the most popular types
of resin have a nominal 8% DVB content.

7. Mobile Ions in Ion-Exchangers

If a cation exchanger contains functional groups in
their original acidic form (e.g., sulfonic acid group),
it is said to be in the H^+ form. When these protons have
been exchanged for other cations, the resin is in the
salt form (e.g., the Na^+ or NH_4^+ form). Similarly, anion
exchangers can be considered to be originally in the OH^-
form, but they are usually converted into other anionic
forms (e.g., the Cl^- form). The Cl^- form is highly
stable; therefore, most anion-exchange resins are shipped
in this form.

The quantity of mobile ions that a resin is capable of
binding is called its capacity; thus the capacity is a
measure of the number of available functional groups in
the resin. The capacity is usually expressed as milli-
equivalents per gram of dry resin in the H^+ or Cl^- form.
For weakly acidic or weakly basic exchangers, the capacity is markedly dependent on pH. This can be shown by
the titration curves of the exchangers (Figure 8.2). The
strongly acidic or strongly basic resins have a much
greater pH range of maximum capacity and are, therefore,
useful over a very wide range.

Although ion-exchange resins generally have satisfactory chemical stability, the temperatures at which they
may be effectively used are limited. For example, the
sulfonic acid resins in the H^+ form react slowly with
water above 100°C to lose sulfuric acid; however, their

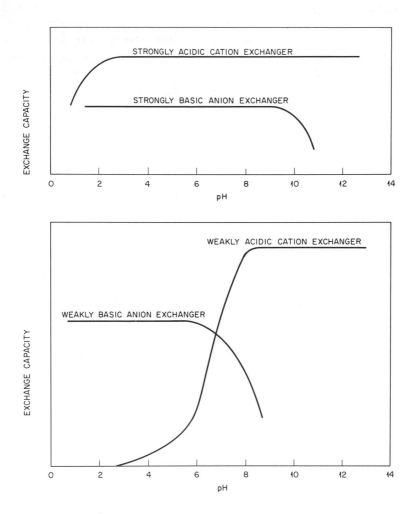

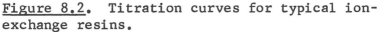

<u>Figure 8.2.</u> Titration curves for typical ion-
exchange resins.

salt forms are sometimes stable up to 200°C for limited
periods of time. When anion-exchange resins containing
quaternary amine groups are in the hydroxide form, they
decompose appreciably above 50°C; however, their salt
forms may be stable up to 100°C. For long-term use of

anion-exchange resins an upper temperature limit of 80°C should be observed.

C. ION-EXCHANGE EQUILIBRIA

To demonstrate the principle of the ion-exchange process in chromatography, consider the transit of a cationic solute through a cation-exchange resin column during a chromatographic run. The sample front advances to the first portion of the resin bed; then, since the solute concentration is at a maximum, the solute ion will tend to exchange with the mobile cation in the resin as follows:

$$R\text{-}H^+ + ION^+ \rightleftharpoons R\text{-}ION^+ + H^+$$

When the solute concentration in the mobile phase decreases at that portion of the bed because of eluent flow and solute sorption, the solute ion will tend to be desorbed and the reverse of the above reaction will occur. Notice that this desorption restores the resin to its original form. As the solute traverses the ion-exchange column, it is successively sorbed and then desorbed. The differences in the degree of sorption and the speed of the sorption step constitute the basis for the chromatographic separation.

1. Distribution Coefficient

In general, ion-exchange methods are based on the distribution of the ionic species of interest between an external solution and the solid phase of the resin. Consider the distribution of an ionic species, A, between the external solution and the ion-exchange resin. Although the equilibrium can be expressed in terms of the mass law, it is usually designated by a distribution coefficient:

$$K = \frac{\text{moles A per gram of exchanger}}{\text{moles A per milliliter of solution}}$$

The ratio of the distribution coefficients of two substances, A and B, under the same set of conditions is frequently termed the separation factor, which is defined as follows:

$$\alpha_B^A = K^A/K^B$$

If α_B^A is greater than 1, A is selectively sorbed and will be retained in the ion-exchange column longer than B. If the separation factor is less than 1, then B is preferred by the resin. This separation factor (or other similar parameters) provides a basis for evaluating the potential resolving power of the resin.

The ion-exchange resins are unique in that they possess a wide range of K values for various ionic species. In some complex mixtures, the K values for the constituents may vary by several orders of magnitude. This extensive coverage of K values (and thus the general applicability to many solutes) is one of the chief advantages of ion-exchange resins; however, it can also necessitate gradient elution chromatography, in which the properties of the eluent are adjusted during the run in order to reduce the high K values of some of the solutes.

In ion-exchange separations, the distribution coefficient can be varied in several ways. It is affected by the concentration of a competing ion in the mobile phase. For example, in a cation-exchange process, an increase in the concentration of an inorganic cation in the mobile phase will decrease the affinity of an organic solute for the resin phase and cause its passage through the column to be accelerated.

Another method used to vary the distribution coefficient is changing the degree of ionization of the mobile ions in the resin. For example, with the weakly acidic and weakly basic exchange resins, the active groups will be un-ionized at basic and acid pH's, respectively. Thus, bound solutes may be eluted from such resins by varying the pH of the eluent. This effect is not as pronounced with the strongly acidic and strongly basic resins. The degree of ionization of the various solutes occurring as a result of the variation of the pH, ionic strength, etc., also has an important effect on the distribution coefficient.

In addition, the distribution coefficient is affected by the operating temperature. Usually, an increase in the temperature reduces distribution coefficients and tends to accelerate the passage of a solute through the column. In some cases, the order of elution of two resolved solutes can be modified or reversed by a relatively small temperature change.

2. Gradient Elution

The time required for a complex separation can be decreased by changing the properties of the mobile phase with time so that the distribution coefficients of the more strongly sorbed species can be reduced during the later stages of the separation. This will allow the entire separative power of the column to be used for the weakly sorbed species during the initial part of the run; then, later in the run, the more strongly sorbed ions can be desorbed more rapidly. This type of chromatography is called gradient elution chromatography.

Eluent properties that can be changed for gradient elution chromatography include pH, temperature, and solvent composition. Changing the solvent composition, usually by increasing the concentration of a competing ion, is probably used more than the other methods. However, in some cases all of these properties are varied with time to achieve a high-resolution, rapid separation.

3. Solvent Uptake

When a dry polystyrene ion-exchange resin is exposed to moisture, it rapidly absorbs an amount of water that is inversely proportional to the degree of cross-linkage. This causes the resin bead to increase in size or to swell. In fact, the amount of absorbed water and the amount of swelling exhibited by the resin bead can be used as indirect measures of the degree of cross-linkage of the resin.

In chromatographic work, the maximum water uptake and the corresponding swollen volume are of primary interest, since they provide a measure of the volume of the resin when immersed in pure water or in very dilute solutions. As the concentration of a solute is increased, the water activity decreases and the resin shrinks. This may be an

important factor in chromatography since some chromatographic separations are made by using an eluent the salt
content of which is varied during the course of the run.
In such cases, the change in the resin dimensions during
the run might be detrimental to separation. This variability in dimensions is much greater in resins with a
lower degree of cross-linkage; therefore, gradient elution chromatography must be carried out with resin of
reasonably high cross-linkage, that is, 8% or greater.

D.   CHROMATOGRAPHY WITH ION-EXCHANGE RESINS

In ion-exchange chromatography, the sample mixture is
first introduced at the top of the ion-exchange column.
Less than 5% of the exchange capacity of the total amount
of resin should be loaded by the sample. Then the absorbed ions are gradually moved down the column by a suitable eluent in a series of sorption-desorption steps.
The eluent either is allowed to flow by gravity to the
column or is forced through the column by a pump or some
other suitable device. If the distribution coefficients
of the constituent ions are sufficiently different, the
species that are being separated will travel down the
column at different rates and will emerge as separate
bands. When one measures a physical property equivalent
to concentrations of the solutes in the column eluate
with time, the resulting histogram will have a series of
peaks, each of which represents one of the solutes.

In general, the separation of different ions of similar size and charge is achieved by making use of the differences in the exchange potentials of the ions. Separation can also depend on the pH and ionic strength of the
sample solution, the operating temperature, and the specific nature of the ion exchanger.

1.   Kinetics of the Ion-Exchange Process

The mechanism affecting the kinetics of exchange in
liquid column chromatography can be described by considering the possible resistances to the exchange of the
solutes between the mobile and the stationary phases.
Five such resistances or mechanisms can affect the rate

process: (1) mass transport of the solute from the mobile phase to the surface of the stationary phase, (2)
mass transport of the solute through the solid phase,
(3) reaction or sorption at the sorption site, (4) mass
transport of the solute back through the solids phase,
and (5) mass transport of the solute from the stationary-
phase surface to the mobile phase. The exchange process
in ion-exchange chromatography can be affected by all of
these mechanisms; however, the mass transport steps are
the most important, and usually the rate of mass transport through the solid phase becomes the predominant
rate mechanism.

Much work has been directed toward developing ion-
exchange materials in which the mass transport resistance
is reduced. This effort has resulted in the manufacture
of very small resin beads and in the development of pellicular resins and macroreticular resins.

2.   Choice of Ion-Exchange Resin

In most instances, satisfactory results can be obtained
with one of the more common types of resin, such as the
sulfonic acid cation-exchange resin or a quaternary ammonium anion-exchange resin. The weakly acidic and weakly
basic resins are rarely used in inorganic chromatography;
however, they have demonstrated some utility in the separation of various organic compounds.

The strongly acidic cation exchangers adsorb all cations that can penetrate the gel matrix. This includes the
adsorption of cations from the salts of strong and weak
acids, as well as the cations of strong and weak bases.
Such resins can be used effectively in elution chromatography in either the H^+ or the salt form. Strongly basic
anion exchangers can adsorb anions from the salts of
strong and weak bases; they can remove the anions from
strong and weak acids; and they can be used in various
forms for the elution chromatography of anions.

Weakly acid cation exchangers do not remove the cations
from solutions of the salts of strong acids; they adsorb
the cations from solutions of strong and moderately
strong bases; and they are satisfactory for use with

alkaline or neutral solutions. Weakly basic anion exchang
ers do not remove the anions from solutions of the salts
of strong bases; they adsorb anions from solutions of
strong and moderately strong acid; they remove anions of
the salts formed from weak bases; and they are satisfactory for use with neutral and acid media.

The properties of the individual solutes in a sample
mixture will usually dictate the choice of ion-exchange
resin. When the properties of the solution are modified
by varying the pH or other parameters, the resulting chemical system may be sufficiently changed to require a different ion-exchange medium. Rather than trying to establish rules in regard to resin choice, several examples of
chromatographic separations utilizing various media, systems and methods have been included in Section F.

3. Choice of Solvent

The majority of chromatographic separations using ion-
exchange resins have been carried out with aqueous solutions because of the superior solvent and ionizing properties of water. However, the usefulness of mixed solvents
such as water/methanol has been demonstrated. It is frequently necessary to use competing solvent ions for optimum elution and at the same time a constant solvent pH.
Various concentrations of aqueous buffers, such as an
ammonium acetate/acetic acid buffer, can be employed for
this purpose.

4. Column Geometry

Column geometry has a significant effect on the resolution or separation achieved in an ion-exchange column.
An increase in the length of the column tends to give
more effective resolution of two components, with an
accompanying increase in peak width. The diameter of
the column should not have a great effect on resolution
(assuming that comparable flow velocities and a propor-
tionally scaled sample size are used) as long as the column is not wide enough to permit radial variations in
fluid properties and not narrow enough to require a sample
size so small that the separated solutes cannot be detected by the column-monitoring system. Column diameters

of 0.1-1.0 cm cover the range of currently usable analytical systems. Some preparative systems use larger columns with an accepted loss of resolution.

5. Resin Bead Size

It is desirable to use resin beads that are spherical, have a narrow size range, and are of the smallest size compatible with the design of the chromatographic system. Very small resin beads have a low resistance to solid-phase mass transport and thus allow a close approach to true equilibrium. Many analytical ion-exchange systems now employ resin particles with diameters less than 40 μ, while the use of resin in the size range of 10 μ (6,7) is becoming commonplace. A diameter of about 2 or 3 μ probably represents the lower limit of useful particle size, since smaller particles are colloidal suspensions.

6. Utility

Although the various types of liquid chromatography have overlapping areas of application, ion-exchange chromatography represents the easiest and most efficient means for separating ionic solutes and is usually the choice when an aqueous solvent is to be used. This is especially important in many biochemical problems because the use of an aqueous chromatographic system simplifies sample preparation. Ion-exchange chromatography is also the logical choice for analyzing many extremely complex sample mixtures since the specificity of many ion-exchange materials allows the separation of literally hundreds of compounds of many different types on one column (5-8).

7. Problems with Ion-Exchange Chromatography

There are, of course, also disadvantages associated with the use of ion-exchange chromatography. For example, it is usually not the best technique for separating solutes dissolved in an organic solvent. Probably an even greater disadvantage is inherent in liquid chromatography in general; that is, mass transport in liquid systems is relatively slow and is usually the controlling resistance to the sorption process. This is further compounded by

the extremely slow mass transport within the resin particle itself. Thus, the ion-exchange process must, of
necessity, be slower than other chromatographic methods,
such as gas chromatography. On the other hand, recent
work on pellicular and macroreticular resins has resulted
in reduced mass transport effects, which will allow some
of these problems to be circumvented (21).

Other problems, which are more or less prevalent in
all separations systems, include the unavailability of
reproducible ion-exchange resins; the unknown sorption
mechanisms associated with many separations (which necessitate an empirical approach to the design of systems);
and the unavailability of suitable, reliable, and cheap
commercial equipment that will allow widespread and routine use of these techniques.

E.   RESIN AVAILABILITY AND TREATMENT

Synthetic ion-exchange resins have been manufactured
for some time in large quantities by several mnufactur-
ers.* In large part, these resins have been used in
water purification applications. In recent years, several other companies** have refined such commercial resins for use with chromatographic systems. Some of these
companies have also developed ion-exchange manufacturing
capabilities of their own (Table 8.1). Extensive lists
of ion-exchange manufacturers and resin types are available from several of the references given in the bibliography at the end of this chapter. Additional information
can be obtained by writing to the various manufacturers.

1.   Treatment of Ion-Exchange Resin

Analytical-grade resin is required for chromatographic
separations. If resin of the required grade or purity is
not available from the manufacturer, commercial-grade
resin can be obtained and further treated to remove

---

* For example, the Dow Chemical Co., Rohm and Haas Co.,
and Pfaudler Permutit, Inc.
**For example, Bio-Rad Laboratories and the J. T. Baker
Chemical Co.

residual materials. This treatment includes extensive washing with an organic solvent (methanol or ethanol), an acidic solution, and a basic solution.

The washing operations for resin composed of relatively large particles (greater than 100 μ in diameter) can be performed by packing the resin into a column and passing a stream of the wash solution through the packed bed until there is no detectable change in the column eluate. Smaller resin should be processed by successively mixing the resin with the wash solutions to form a slurry and then filtering and washing with distilled water between treatment steps.

Cation-exchange resins are usually supplied in the H^+ or Na^+ form, and anion-exchange resins in the Cl^- form. Occasionally, however, it is desirable to modify these resins for use in a different ionic form. In general, the ionic form of a particular resin can be changed by placing the resin in contact with an aqueous solution containing a high concentration of the desired cation or anion. Again, depending on particle size, this can be done by packing the resin into a column and percolating the reagent through it until conversion is complete, or by successively mixing a slurry of the resin with a solution of the new ion followed by filtration and washing with distilled water.

To convert an anion-exchange resin from a form with strongly bound anions (e.g., the Cl^- form) to a weak anionic form, such as acetate or formate, it is more convenient to first convert the Cl^- resin to the hydroxide form and then bring the subsequently obtained OH^- resin into contact with the desired anion.

2. Variations in Ion-Exchange Resin

Batch-to-batch variations in the stationary phase in ion-exchange chromatography can cause difficulties. Such variations occur in ion-exchange resins because of the numerous important parameters that must be considered during the manufacture of these resins. Several years ago, a worker in ion-exchange chromatography often obtained a small amount of a resin that enabled him to

achieve a complicated or unusual separation; later, however, he would experience the frustration of finding that
no more of the same batch of resin was available and that
no other resin would duplicate his earlier results. The
situation has improved somewhat now, although variations
in resin from batch to batch are still occasionally encountered.

Obviously, a means for determining the significant
properties of the resin is needed. Also, a method for
relating the physical properties (e.g., particle size,
type of active group, and degree of cross-linkage) of the
resin to its separative characteristics would be very useful. Some interesting results in this regard have been
reported by David Freeman and his co-workers at the National Bureau of Standards (9).

3. Particle Size of the Resin

One of the most important physical properties of an
ion-exchange resin is the size of its individual particles. Most high-resolution systems currently use spher-
ically shaped resin having an average particle size less
than 20 µ in diameter. It is important for this resin to
be uniform in size (i.e., exist in a narrow size range)
since this uniformity enhances the flow properties and
reduces the pressure drop in the column.

Resin having a particle size greater than about 40 µ
in diameter can be separated into narrow size ranges by
using mechanical sieves. However, water elutriation
(10) is a more effective way to separate particles smaller than 40 µ in diameter (Figure 8.3). In water elutriation, the upward flow of a liquid (usually water) in a
column subjects the resin mixture, which is introduced
as a slurry, to a flow field that tends to carry over the
small particles while allowing the larger ones to fall
into a bottom collection chamber. By using this technique, the resin can be separated into a particle size
range as narrow as ±1 µ.

Until recently, the smallest spherical ion-exchange
resin available was a nominal minus 400-mesh fraction containing particles in the 5-60 µ diameter range (Figure

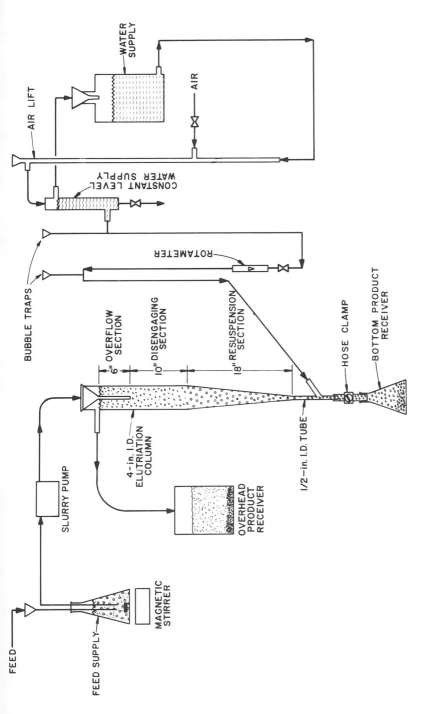

Figure 8.3. Continuous elutriation system for separating ion-exchange resin into size fractions. Reprinted from ref. 10 by permission of Analytical Biochemistry.

305

8.4). To obtain the small resin, large quantities of the

FEED MATERIAL: NOMINAL MINUS 400 MESH

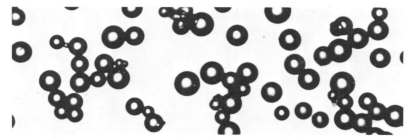

NOMINAL 10 TO 20-MICRON-DIAMETER RESIN

NOMINAL MINUS 10-MICRON-DIAMETER RESIN

0   50
MICRONS

<u>Figure 8.4.</u> Commercial-grade Dowex 1-X8 ion-ex-
change resin separated into two size fractions by
water elutriation.

resin had to be elutriated. Now, however, a few of the resin suppliers* have installed facilities to provide such separations as a routine service; thus the resin can be purchased, already prepared in the desired size range.

4. Choice of Ion-Exchange Columns

An ordinary laboratory burette can be used as the column in ion-exchange separations when automation of an analytical method is not desired and relatively large resin particles will be used. A plug of cotton or glass wool will serve to support the resin. The diameter of such columns should not exceed 10 mm; and, if gravity flow is to be used, the length of the column should not be greater than 50 cm (Figure 8.5).

Most ion-exchange techniques used today employ forced flow in which the eluent is supplied to the column by a pump. If the pressure drop across the resin bed is expected to be appreciable, great care must be exercised in the choice of the chromatographic system. Several chromatography suppliers** now stock ion-exchange columns that are equipped with a suitable porous support in the bottom and a fitting on each end for supplying the eluent and for withdrawing the eluate, respectively.

Glass columns that can withstand pressures as high as 1000 psi are available. These columns have internal diameters as large as 0.9 cm and are up to 150 cm long. Metal columns, which can be easily fabricated from seamless metal tubing (Figure 8.5), are used for high-pressure techniques (i.e., those requiring pressures greater than 1000 psi). Conventional compression tubing fittings can be employed for the fluid entrance and exit, and for holding a porous metal support for the resin bed.

---

*For example, Bio-Rad Laboratories and Durram Instrument Co.
**For example, Chromatronix, Inc., and Metaloglass, Inc.

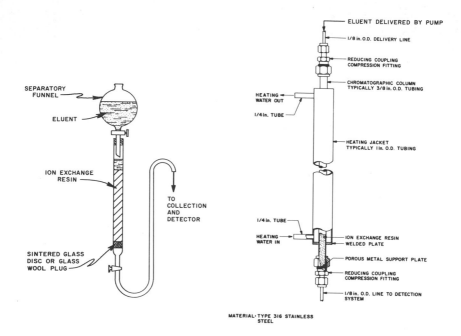

SIMPLE LABORATORY ION EXCHANGE COLUMN FOR GRAVITY FLOW   HIGH-PRESSURE CHROMATOGRAPHIC COLUMN FABRICATED FROM STAINLESS STEEL TUBING.

<u>Figure 8.5.</u> Columns that can be used in ion-exchange chromatography.

5. <u>Procedure for Packing Columns with Resin</u>

Although the packing of a chromatographic column with ion-exchange resin requires less skill than is needed for other types of stationary phases, it should be done with a reasonable amount of care. The conventional method for packing a column with relatively large, closely sized ion-exchange resin is to (1) make a slurry of the resin particles, (2) introduce the slurry into the column or a chamber attached to the column, and (3) allow the resin particles in the slurry to settle by gravity. The supernate is removed when necessary, and additional slurry is added until the column is completely filled with resin.

This technique has distinct disadvantages when the size of the resin particles varies over a large range. The larger particles will settle more rapidly, and the resulting bed will be composed of longitudinal zones containing particles of various sizes. Also, if the column is not held exactly vertical during the packing procedure, radial variation in particle size will occur. Such variations affect the resolution of the column and make it difficult to pack several columns having reproducible characteristics. An additional problem exists when a chromatographic column is being packed with extremely small resin particles (less than 20 μ in diameter). With the gravity method of packing, the settling velocity of these particles is very low and the time required to prepare a given column is prohibitive. In such cases, the technique of dynamic packing has been successfully used (11).

In dynamic packing, the ion-exchange resin particles, which are contained in a flowing fluid, are forced into the packed bed at a velocity much greater than their settling velocity. Dynamic packing can be done in either of two ways: (1) by displacing a thick slurry, or (2) by extruding a prepacked bed. Packing via slurry displacement is accomplished by connecting a chamber or reservoir to the chromatographic column, filling the chamber with a thick slurry of the ion-exchange resin, and then displacing the slurry into the column with a liquid that is pumped into the top of the slurry chamber (see Figure 8.6). If the linear velocity of the displacement fluid in the slurry chamber is substantially greater than the settling velocity of the largest particle, size segregation will not occur in the resulting packed bed. The resin should be packed with a liquid flow rate greater than the anticipated eluent flow rate for the chromatographic run.

When small-diameter columns of considerable length are packed with finely divided ion-exchange resin, difficulty is encountered in pumping liquid through the slurry chamber and into the column at a sufficiently

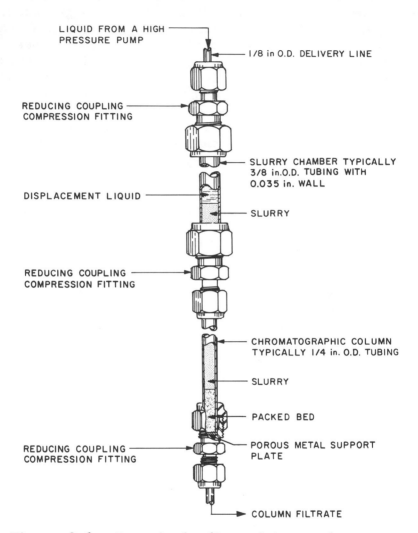

Figure 8.6. Dynamic loading of ion-exchange resin into chromatographic columns.

rapid rate to prevent size segregation. This difficulty is caused by a high pressure drop (in the long column), which prevents rapid displacement of the slurry from the slurry chamber into the column. However, as an alternative, a fixed bed can be packed into a reservoir, or cartridge of larger diameter (than the column), and then extruded into the small-diameter chromatographic column by displacement with liquid.

This extrusion technique can also be used to pack a coiled column with resulting constant packing properties. In addition, it has proved to be useful for preparing column charges in a cartridge to be transported from one laboratory to another. Such a method might be exploited by resin manufacturers as a means of providing customers with new column charges, which would allow the user to pack a column with minimal effort.

6. Equilibration

After a column has been packed with ion-exchange resin, the resin must be equilibrated with the flowing eluent before a chromatographic separation is made. This equilibration procedure may require a period of several hours, especially if the resin was previously in equilibrium with a solution having a much different ionic strength from that of the eluent.

Ion-exchange resin can be used many times for the same separation before its exchange capacity is exhausted. If operating conditions are constant during an entire run, the column is ready for the next run as soon as the last constituent is eluted. However, if gradient elution or temperature programming is used during the run, the resin must be re-equilibrated at the starting conditions before it can serve for the succeeding run.

7. Life of the Column

A resin life of 6 months is not unusual in analytical systems. This means that hundreds of simple chromatographic separations can be made on a single column without replacing the resin. For separations in which some material is irreversibly sorbed (e.g., in some biochemical systems), a small "precolumn" or ion-exchange cartridge placed directly ahead of the ion-exchange column is effective in trapping the material. This precolumn can be routinely replaced after a few runs.

8. Introduction of the Sample

The sample, or feed solution, can be introduced to the column in two ways, depending on the type of separation desired. In a preparative or production run, in which a large volume of solution must be separated and in which the components of the mixture adsorb satisfactorily, loading can be achieved by simply pumping the feed solution into the chromatographic column. After this step is complete, a different eluent is used to achieve separation and to elute the separated components.

In most analytical ion-exchange chromatographic systems, a small amount of the sample mixture is placed directly on the column or into the eluent stream. A simple, inexpensive way of doing this is to temporarily stop the eluent flow, remove a small amount of eluent from the top of the column, inject the sample onto the top of the column, and then force the sample to move down the resin, either by gravity flow or by air pressure. After the void space in the column has been replaced with eluent, the run is started. Such a technique is time-consuming, however, and subject to operator error.

A more effective method is to introduce the sample directly into the eluent just before it comes into contact with the ion-exchange resin. This can be done by injecting the sample via a hypodermic syringe through a septum into the top of the column. Although this syringe technique is satisfactory with low-pressure systems, it cannot be used at pressures greater than about 1000 psi. At higher pressures the sample should be introduced by means of a sample injection valve (Figure 8.7). Such valves are usually constructed with six ports, each pair being interconnected. In one orientation, a sample can be loaded into a sample loop; and, when the ports are reoriented by turning the valve handle, the sample loop is made part of the eluent line. Valves that allow automated sample introduction at pressures up to 5000 psi without interrupting the system have now been developed (12).

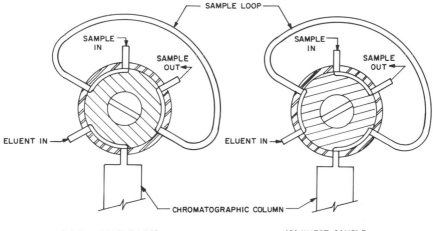

Figure 8.7. Use of a six-port valve to inject a sample into the eluent stream of a chromatograph.

F. EXAMPLES OF ION-EXCHANGE CHROMATOGRAPHY

Since ion-exchange chromatography has been developed largely in an empirical manner, one of the best ways to discuss the choice of resin, design of systems, operational methods, etc., is by citing several successful examples. Many such applications will be presented in Part III of the book, but a few of historical significance will be mentioned here. Methods using high-pressure ion-exchange chromatography will be discussed in somewhat more detail since that technology apparently will be very important in the future.

1. Inorganic Separations

The utility of liquid chromatography was first demonstrated in separations of fission products. Thereafter, the striking success of such workers as Tompkins (2) and Ketelle and Boyd (3) in separating rare earths by ion-exchange chromatography directed the attention of researchers in various fields to the potentialities of chromatographic methods.

One use of anion-exchange chromatography is the rapid analysis of phosphates in detergents. By using this method, the ortho- and pyrophosphates and the tripoly- and trimetaforms can be separated and quantified on Dowex 1-X8 resin (13). Cation-exchange chromatography is a widely accepted method for separating alkali metals by elution with an aqueous mixture of HCl and ethanol (14).

The transition elements manganese through zinc can be separated on Dowex 1-X10 by converting the metal ions to their anionic chloride complexes in high concentrations of HCl (15). This separation demonstrates the significant change in distribution coefficient that can be achieved by adjusting the properties of the eluent (i.e., changing the sorbed solute from a cation to an anion form).

A recent example of high-pressure ion-exchange chromatography used for the separation of inorganic compounds is the separation, on a production scale, of synthetically produced transplutonium elements. Ion-exchange techniques for separating lanthanides and actinides have been adapted for this separation. The technique involves loading the trivalent lanthanide and actinide elements, as cations in a weakly acid solution, onto a cation-exchange column and then eluting them sequentially with an anionic complexing agent (16).

When the elements to be separated are highly radioactive (e.g., americium, curium, berkelium, californium, and einsteinium), difficulty has been encountered with ion-exchange resin because of radiation-induced resin damage and radiolytic gas evolution. Both of these problems, which effectively destroy the resolving ability of the chromatographic column, can be circumvented by using small (10-20 μ in diameter) resin particles. This finely divided resin will permit satisfactory separation in a much shorter time, thus reducing the exposure of the resin to the radioactive materials. Also, the resulting high-pressure operation (up to 2000 psi) increases the solubility of the radiolytic gases and thus prevents the deleterious effects of gas bubbles (17).

Figure 8.8 shows a typical chromatogram obtained for this system. Scaled-up versions are now being used to separate entire production runs of these elements in a procedure that requires about 2 hr.

2. Biochemical Separation

Ion-exchange chromatography was first applied to biochemical problems by Cohn (4) and his collaborators, who used it to separate nucleic acid derivatives, and Moore and Stein (5), who separated amino acids.

Recent developments in this field have been directed toward (1) decreasing the analysis time for the separation of the more simple biochemical systems, and (2) developing separation systems for the analysis of very complex biochemical mixtures, notably physiologic fluids.

For example, the time for the chromatographic separation of the 18 amino acids commonly occurring in proteins was reduced to about 22 hr, using a two-column system with Dowex 50 cation-exchange resin and gradient elution with a citrate buffer (18). This procedure has been improved by using new resins of increasingly smaller diameter until an analysis time of about 2 hr is now possible (19).

At the same time, technology is being developed for the analysis of amino acids and other ninhydrin-positive compounds in physiologic fluids. Hamilton (8) has developed high-resolution systems for such fluids and has separated more than 175 ninhydrin-positive chromatographic peaks from a single urine sample in an analysis time of over 2 days.

An ion-exchange separation of the principal purine and purimidine bases which occur in nucleic acid mixtures was first made by Cohn (4), who exploited the cationic properties of these compounds in an acid solution. Dowex 50 resin was used to pack the column, and $2\underline{N}$ HCl was used as the eluent. Separation required about 16 hr. Cohn also separated the corresponding nucleosides by cation-exchange chromatography. This work has been continued by Uziel, Koh, and Cohn (20), who have succeeded in reducing the analysis time for common ribonucleosides to approximately 1 hr.

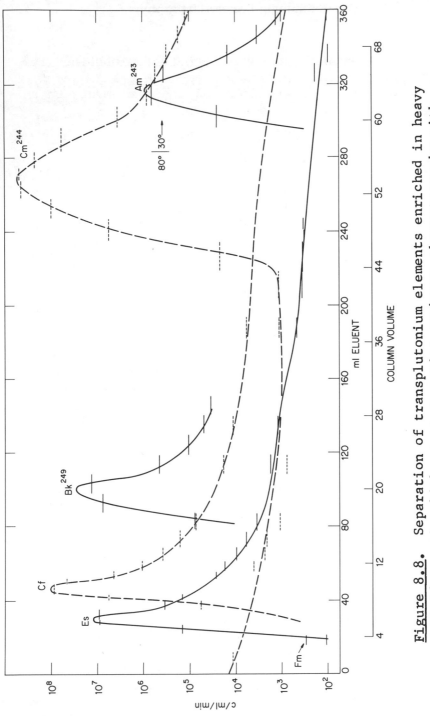

Figure 8.8. Separation of transplutonium elements enriched in heavy actinides, using high-pressure cation-exchange chromatography with Dowex®50-X12 in a column 66 cm x 0.17 cm² and an aqueous solution of α-hydroxyisobutyric acid as the eluent. Reprinted from ref. 17 by permission of the American Chemical Society.

3.   High-Pressure Ion-Exchange Chromatography

The requirements for high-resolving separations systems in which finely divided ion-exchange resin is used in long columns have resulted in the development of

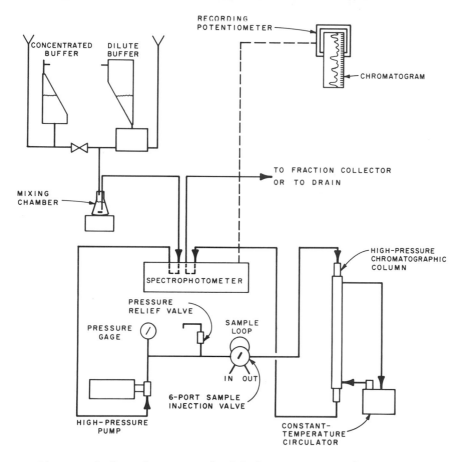

Figure 8.9. Automated, high-pressure chromatograph for analysis of the UV-absorbing constituents in body fluids, using 10-μ-diameter Aminex® A-27, anion-exchange resin in a 150 cm x 0.62 cm stainless steel column with gradient elution of an aqueous acetate buffer.

high-pressure ion-exchange chromatography. Two such

systems for the high-resolution analysis of body fluids
are discussed in order to show the utility of ion-exchange
chromatography for separating very complex mixtures: an
analyzer for the ultraviolet absorbing constituents (Figure 8.9) and an analyzer for the carbohydrates in body

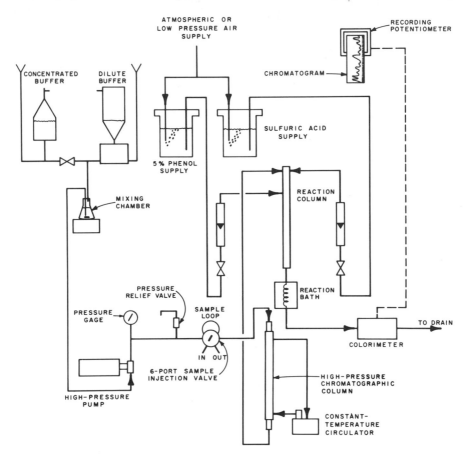

Figure 8.10. Automated, high-pressure chromatograph for analysis of the carbohydrates in body
fluids, using 10-μ-diameter Aminex® A-27, anion-exchange resin in a 150 cm x 0.62 cm stainless
steel column with gradient elution of an aqueous
borate buffer. Reprinted from ref. 7 by permission
of American Journal of Clinical Pathology.

fluids (Figure 8.10). Prototypes of both analyzers have
been designed and built and are currently being tested in
clinical and medical research laboratories (7).

Both analyzers feature (1) heated, high-pressure (up
to 5000 psi) anion-exchange columns packed with nominal
10-μ-diameter quaternary ammonium resin (Bio-Rad Amenix®
A-27); (2) gradient elution with an aqueous buffer to
separate and transport the separated constituents of the
sample mixture; and (3) a recording photometer or color-
imeter for the detection and quantification of the separated constituents.

Samples are introduced into the system by injecting a
measured volume (0.5-2.0 ml) of fluid by means of a six-
port injection valve, into the high-pressure eluent
stream just ahead of the ion-exchange column. Analytical
results are presented graphically in the form of a chromatogram showing the absorbance of the eluate stream versus time. Each detectable molecular constituent is represented by a chromatographic peak. In the typical urine
chromatogram, 100-120 UV-absorbing peaks are resolved in
about 40 hr (Figure 8.11); a maximum of 150 peaks has
been observed. As many as 48 chromatographic peaks have
been resolved from a single urine sample on the carbohydrate analyzer (Figure 8.12).

G. THE FUTURE OF ION-EXCHANGE CHROMATOGRAPHY

The future for ion-exchange chromatography looks promising. As manufacturing techniques are improved, ion-
exchange resins will undoubtedly exhibit reproducibility
from batch to batch. New types of ion-exchange resins
with different functional groups will become available.
The new pellicular resins and other new physical forms of
resin will have a great impact on high-speed liquid chromatography (21). Examples of the use of some of these
materials are given in Part III of this book.

As reproducible resins become available, the use of
different types of resins in sequence will become possible. For example, an anion-exchange resin column in
sequence with a cation-exchange column could serve to
separate an extremely complex mixture; after the initial

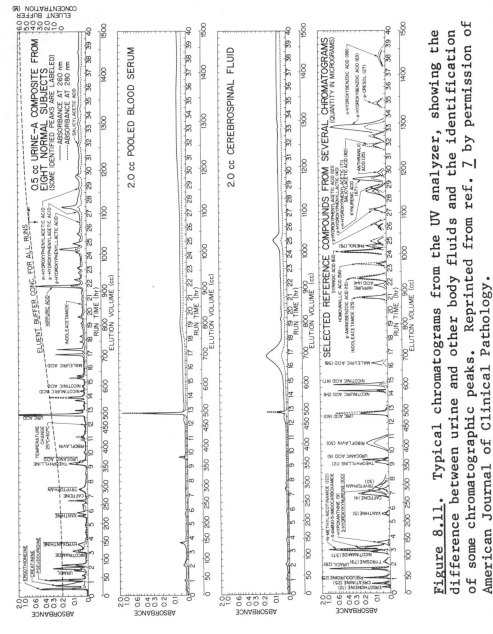

Figure 8.11. Typical chromatograms from the UV analyzer, showing the difference between urine and other body fluids and the identification of some chromatographic peaks. Reprinted from ref. 7 by permission of American Journal of Clinical Pathology.

320

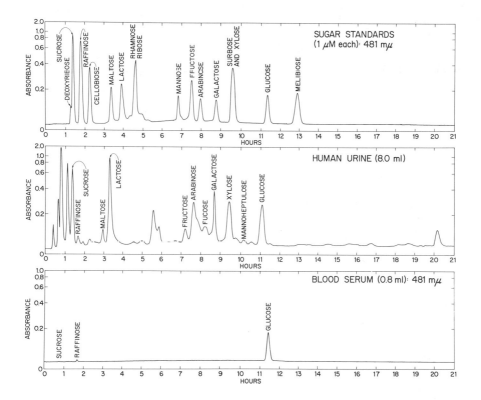

<u>Figure 8.12.</u> Typical chromatograms from the car-
bohydrate analyzer, showing the difference between
urine and blood serum and the identification of
some of the chromatographic peaks.

loading step, the two columns could be eluted separately
at the same time to reduce the total separation time.

New uses for ion-exchange chromatography will undoubtedly be found in the high-resolution separation of complex mixtures. Since many such mixtures are important
in biology and medicine, these areas will undoubtedly be
fertile fields of application for ion-exchange chromatography.

Finally, before ion-exchange chromatography and, in
particular, high-resolution chromatography that

necessitates high-pressure operation become generally useful, the instrument companies must produce efficient, automated liquid chromatographs at a reasonable price.

Selected Bibliography

Fundamentals of Ion Exchange

Kunin, R., "Ion Exchange Resins," 2nd ed., John Wiley, New York, 1958.

Helfferich, F., "Ion Exchange," McGraw-Hill, New York, 1962.

Marinsky, J. A., "Ion Exchange: A Series of Advances," Vol. 1, Marcel Dekker, New York, 1966.

Ion-Exchange Chromatography

Miles, O., ed., "Laboratory Handbook of Chromatographic Methods," Van Nostrand, Princeton, N. J., 1961.

Samuelson, O., "Ion Exchange Separations in Analytical Chemistry," John Wiley, New York, 1963.

Morris, C. J. O. R., and P. Morris, "Separation Methods in Biochemistry," Wiley-Interscience, New York, 1964.

Heftmann, E., ed., "Chromatography," Reinhold, New York, 1964.

Giddings, J. Calvin, and Roy A. Keller, eds., "Advances in Chromatography," Vols. 1-6, Marcel Dekker, New York, 1965.

Inczedy, J., "Analytical Applications of Ion Exchangeers," Pergamon, London, 1966.

Dean, J. A., "Chemical Separation Methods," Van Nostrand, Princeton, N. J., 1969.

References

1. B. A. Adams and E. L. Holmes, J. Soc. Chem. Ind., 54T, 1 (1935).
2. E. R. Tompkins, J. X. Khyn, and W. E. Cohn, J. Am. Chem. Soc., 69, 2769 (1947).
3. B. H. Ketelle and G. E. Boyd, J. Am. Chem. Soc., 69, 2800 (1947).
4. W. E. Cohn, Science, 109, 377 (1949).
5. S. Moore and W. H. Stein, J. Biol. Chem., 192, 663 (1951).

6. C. D. Scott, Clin. Chem., $\underline{14}$, 521 (1968).
7. C. D. Scott, R. L. Jolley, W. W. Pitt, and W. F. Johnson, Am. J. Clin. Pathol., $\underline{53}$, 701 (1970).
8. P. B. Hamilton, "Handbook of Biochemistry: Selected Data for Molecular Biology," The Chemical Rubber Co., Cleveland, 1968, pp. B-43 to B-55.
9. D. H. Freeman, V. C. Patel, and M. E. Smith, J. Polymer Sci., $\underline{3}$, 2893 (1965).
10. C. D. Scott, Anal. Biochem., $\underline{24}$, 292 (1968).
11. C. D. Scott and N. E. Lee, J. Chromatog., $\underline{42}$, 263 (1969).
12. C. D. Scott, W. F. Johnson, and V. E. Walker, Anal. Biochem., $\underline{32}$, 182 (1969).
13. D. P. Lundgren and N. P. Loeb, Anal. Chem., $\underline{33}$, 366 (1961).
14. F. W. E. Strelow, C. J. Liebenberg, and F. Von S. toerien, Anal. Chim. Acta, $\underline{43}$, 465 (1968).
15. K. A. Kraus and F. Nelson, ASTM Spec. Publ. No. $\underline{195}$ (1958), p. 27.
16. D. O. Campbell and S. R. Buxton, Ind. Eng. Chem. Process Design Develop., $\underline{9}$, 89 (1970).
17. D. O. Campbell, Ind. Eng. Chem. Process Design Develop., $\underline{9}$, 95 (1970).
18. S. Moore, D. H. Spackman, and W. H. Stein, Anal. Chem., $\underline{30}$, 1185 (1958).
19. "Price List U: Ion Exchange Resins and Systems," Bio-Rad Laboratories (1969).
20. M. Uziel, C. K. Koh, and W. E. Cohn, Anal. Biochem., $\underline{25}$, 77 (1968).
21. C. G. Horvath, B. A. Preiss, and S. R. Lipsky, Anal. Chem., $\underline{39}$, 1422 (1967).

CHAPTER 9

Liquid Chromatography ... An Overview

István Halász

A. INTRODUCTION

The linear velocity of the eluent in classical liquid chromatography, well known and broadly applied since the beginning of this century, is 0.001-0.01 cm/sec. In "high-speed" or "high-pressure" chromatography, developed in the last few years, velocities in the order of 1 cm/sec are used. This velocity is comparable with that achieved in conventional packed columns in gas chromatography. Most of the workers developing high-speed liquid chromatography were outstanding experts in the field of gas chromatography who tried to "translate" and to apply their theoretical knowledge and experimental skill to this new field.

Although the basic theory of chromatography is common, there are (among others) three basic differences in applying this theory: (1) the interdiffusion coefficient of liquids is at least 10^4 times smaller than that of gases, (2) the viscosity of the eluent is roughly 100 times greater for liquids than for gases, and (3) the interaction between the molecules of the stationary phase and those of the eluent are negligible in gas chromatography, but important in liquid chromatography. On the other hand, the theoretical treatment of liquid chromatography is simpler because the mobile phase is not compressible. Unfortunately, in the field of chromatography there has been a gap between the theoretical experts and those who are interested mainly in the separation of given samples. Knowledge of the separation mechanism for a given problem may not be absolutely necessary. However, it must not be forgotten that an understanding of the separation mechanism made it possible to develop better types of columns. Therefore, knowledge of the simple basic theory for routine work is indeed necessary. This pragmatic theory, as presented in several excellent chapters in this book,

provides a starting point for finding optimal separation
conditions without excessive experimental work.

B. SOME EXPERIMENTAL FACTORS AFFECTING SEPARATIONS

1. Interdiffusion Coefficient

Small interdiffusion coefficients results in a slow
speed of mass transfer in the mobile phase. Consequently,
liquid chromatography with open tubes (capillary columns)
results in intolerable peak broadening if the flow is
laminar, as calculated using the equations of Taylor (1),
Aris (2), or Golay (3). To improve mass transfer in the
mobile phase, small particle sizes (d_p = <80 μ) are required, resulting in increased flow resistance of the
packed column. Furthermore, as pointed out by some of
the authors of this book and as described in the literature (4), the packing of columns with particles for which
d_p = <25 μ is difficult and not always reproducible.

2. Viscosity

As a consequence of the high viscosity and small particle sizes, column pressures of 10-100 atm are required
in high-speed liquid chromatography. Pumps with pressures over 500 atm are available, but with such high pressures the overall construction of the equipment becomes
sophisticated. Furthermore, the volume flow rate from a
pump operated at constant conditions decreases with increasing column back-pressure. At this stage of development inlet pressures of about 500 atm seem to be the
upper limit. Consequently, the length of the column which
can be used is limited. In most of the separations published to date, the number of effective plates, N_{eff},
generated in high-speed liquid chromatographic columns is
smaller than 1000 and usually smaller than 500.

3. Interaction Between the Eluent and the Stationary Phase

The sample and the eluent have to be chemically "similar" because the sample must dissolve. On the other
hand, partition of the sample between the mobile and the
stationary phases occurs only if the sample and the

stationary phase are "similar." Consequently, the liquid
stationary phases are usually somewhat soluble in the mobile phase and the column bleeds if the eluent is not
presaturated with the stationary phase. However, exact
presaturation is very difficult, and the gain or loss of
stationary phase due to a lack of equilibration results
in changed retention. These effects become more important at the lower concentrations of stationary phases on
the column. From this point of view, the use of active
solid or chemically bonded stationary phases is preferred.

4. Sample Injection

Three possibilities have been proposed for injecting
the sample against the high inlet pressure of the column:
(1) pressureless injection by interrupting the eluent
flow (5), (2) rotating valve injection (6), and (3) syringe injection into the flowing system (7).

Experimentally, the first method is the simplest.
However, flow rate equilibrium is not instantaneously
achieved after the injection; therefore, the retention
times are not measured precisely. Furthermore, the mechanical degradation of the packing material is increased
by stopping and starting the flow.

Because of capillary forces and mechanical problems of
construction, the smallest sample size that can be used
with a rotating valve is about 3 µl (8).

When the sample is injected with a syringe against
high inlet pressure, the syringe has to be tight enough
and the "fountain" effect must be overcome. The latter
effect occurs when, as the needle is withdrawn, the septum does not tighten and leaks because of the high liquid
pressure. With proper mechanical construction of the
injection system this problem can be solved (7). One
type of syringe (Hamilton Company, Whittier, Calif., 5-µl
high-pressure syringe, Model HP305N) was found to be
tight up to 400 atm. The durability of the septum is determined by the extent of attack from the mobile phase
used.

In view of the dissimilarity of the approaches, it is surprising that peak broadening due to sampling is practically identical with all three techniques (8).

5. Connecting Tubes

Because of the small interdiffusion coefficient of liquids, the connecting tubing (between the injection port and the column and between the column and the detector) and the detector itself (especially with large dead volume) can negate the separation achieved in the column if the flow is laminar (8,9,10,11,12). This loss in resolution is large with an unretained sample and with a nonporous support, and increases with decreasing column internal diameter. It is possible, for example, to observe a false maximum in the curve of plate height versus carrier velocity instead of a linear ascending branch, the latter measured with proper equipment (9). The relative peak broadening may be increased up to a factor of 10 with a connecting pipe only 0.25 mm in internal diameter and 20 cm in length. On the other hand, the equipment often simulates a performance much better than the actual one inside the column (12). Unfortunately, with the apparatus often used in routine high-speed liquid chromatography, one measures anything but the true performance inside the column (12).

As shown by calculation and experiments (12), the measured resolution, R_S, is independent of the volume of pipes and detector per se, that is, if the flow itself causes no broadening in these parts of the equipment. Disturbing the laminar flow pattern by deforming the geometry of the open tube may cause the peak broadening to decrease dramatically, especially at high linear velocities of the eluent (9,11).

6. Retention Time, t_0, of an Inert Peak

In liquid chromatography it is always difficult to determine the true t_0, the hold time of the column, because it is difficult to predict whether or not a sample is inert. There are four different approaches to solving this problem.

(1) Exact t_0 values are achieved by measuring the retention of the radioactively labeled molecules of the
eluent itself. For this, however, a special detector is
needed. Furthermore, the volume of the detector used in
routine work is never identical with that of the special
one, resulting in changed holdup time. (2) A homolog
with a lower carbon number than the eluent is usually unretained. (3) If the corrected retention of a compound
is constant in a temperature range of 20-50°C, it may be
considered an inert peak. (4) With a regular packed
column where the ratio of inner diameter, d_c, to particle
size, d_p, is greater than 10, one can approximate the linear velocity of the eluent. If the support is porous
(e.g., silica, alumina, Chromosorb®, or Porasil®)

$$v = \frac{1.5F}{d_c^2}$$ (9.1)

where F is the volume flow rate (cc/sec), d_c the inner
diameter of the column (cm), and v the linear velocity
of the eluent (cm/sec).

Using liquid-impenetrable, nonporous supports such as
glass beads gives

$$v = \frac{3F}{d_c^2}$$ (9.2)

By dividing the column length, L, by v the retention time
of the inert peak is determined.

7. Efficiency and Permeability

For all types of columns

$$\Delta P = \frac{v\eta L}{K^0}$$ (9.3)

where ΔP is the pressure drop in the column (atm x 10^6),
η the viscosity (poise), and K^0 the permeability (cm^2).
As long as the flow is laminar, K^0 is independent of the
inner diameter of the column, the type of eluent, and the
temperature. If the particle size in a regular packed
column is greater than 20-30 μ, a good approximation is

$$K^o = \frac{d_p^2}{1000} \qquad (9.4)$$

or, from equations 9.3 and 9.4,

$$\Delta P = \frac{10^3 \, v\eta L}{d_p^2} = \frac{10^3 \, v\eta HN}{d_p^2} = \frac{10^3 \, v\eta H_{eff} N_{eff}}{d_p^2} \qquad (9.5)$$

In gas and liquid chromatography, if the particle size is greater than 80-100 μ, on the ascending branch of the H versus v curve, $H \sim d_p^2$, with all other parameters constant. Inserting this assumption in equation 9.5 gives

$$\Delta P \sim 10^3 \, v\eta N \sim 10^3 \, v\eta N_{eff} \qquad (9.6)$$

As seen from equation 9.6, if $H \sim d_p^2$, the number of plates, N or N_{eff}, generated in a column with a given pressure drop, ΔP (at a given linear velocity, v, and with a given viscosity, η, of the eluent), is independent of the particle size of the support. Of course, the column length required to maintain a fixed number of plates increases with increasing particle size in this case. For example, when increasing d_p by a factor of 3, one has to elongate the column by a factor of 9 to generate the same number of plates, N or N_{eff}. The permeability, however, will increase by the same factor of 9, resulting in unchanged pressure drop on the elongated column.

8. Sieve Fraction

Unfortunately, it seems to be an empirical fact that the permeability, K^o, is proportional to the smallest, and the plate height, H_{eff}, to the largest, particle size, d_p, within a given sieve fraction. Consequently, the ratio of the largest and smallest particle sizes must be as small as possible to achieve the wanted minimum H/K^o ratio.

C. SOME CONSIDERATIONS IN COLUMN PERFORMANCE

1. Speed of Analysis

The speed of the analysis is characterized by the number of theoretical, N, or effective, N_{eff}, plates generated per second, where

$$\frac{N_{eff}}{t} = \frac{v}{H_{eff}(1 + k')} \tag{9.7}$$

and H_{eff} is the height equivalent to an effective plate.
As seen from equation 9.7, the speed of analysis always
increases with decreasing H_{eff}, that is, with decreasing
particle size. Consequently, the smallest available particle size must be used to achieve the optimal speed of
analysis. When the particle size is decreased, and the
pressure drop and the velocity are held constant, N_{eff}
will remain constant as long as $H_{eff} \sim d_p^2$.

In liquid chromatography, however, d_p values of 20-70
μ are usual. In this region H_{eff} is proportional, not to
d_p^2, but to $d_p^{1.4}$ to $d_p^{1.8}$ ($\underline{4},\underline{8}$). Sometimes the exponent
of d_p depends on the capacity ratio, k', of the sample,
although all the measurements were made in the same col-
umn ($\underline{8}$). In this region N_{eff} will decrease (because of
the shorter column length), but N_{eff}/t will increase with
decreasing particle size if the velocity of the eluent
and the pressure drop on the column are constant.

To achieve the same value of N_{eff}, velocity has to be
reduced in a column packed with particles of smaller size.
Now, with a given pressure drop, this column can be elongated, resulting in higher N_{eff} values. If the speed of
analysis, N_{eff}/t, is identical in two columns packed with
different sized particles (with different v but identical
ΔP), the column which will have the higher plate number
cannot be predicted. This depends on the quality of the
stationary phase, that is, on the speed of mass transfer
in the stationary phase, and on the exponent β in the
$H_{eff} \sim d_p^\beta$ relationship. Consequently, in the range of
d_p = 20-70 μ the optimum particle size is not predictable
and must be determined by experiment. A choice may then
be made between efficiency and speed.

2. Comparison of Column Efficiency

When comparing, with the help of reduced plate height
data, columns packed with different sieve fractions of
the support, one has to be extremely cautious. Here
H/d_p (reduced plate height) is plotted versus vd_p/D_m

(reduced velocity), where D_m is the interdiffusion coefficient in the mobile phase. This plot, of course, is
equivalent to that of $H/d_p{}^2$ versus v/D_m. The curves will
be identical, that is, independent of particle size, only
if $H \sim d_p{}^2$. This is unusual, however, in high-speed
liquid chromatography. Consequently, in this presentation the curves for columns with smaller d_p always appear
at a disadvantage if the packing (except d_p) is identical.

When reduced plate height data are plotted, d_p is
really undefined for a sieve fraction. It is undefined
whether the smallest, the average, or the largest d_p of
the sieve fraction has to be used for the calculation of
the reduced plate height. This problem becomes extremely
difficult if the ratio of the largest to smallest particle
size of the sieve fractions compared is not large. If an
accurate comparison of the efficiencies of different packing materials is desired, it must be carried out with
identical sieve fractions. From the point of view of
separation, of course, the H_{eff} and not the H versus v
curves have to be compared. Thus, the comparison depends
on the nature of the sample to be resolved.

In the literature, H/d_p values are occasionally compared in a given linear velocity range (or, which is
worse, at constant flow rates) of the eluent. This presentation has no meaning whatever. The reduced plate
heights should be compared at the same reduced velocity,
if at all.

3. Optimum Linear Velocities

From equation 9.7, one can see that the time of analysis is proportional to H_{eff}/v. The price to be paid for
greater speed is a change in ΔP: the pressure drop increases with increasing velocity. Velocities used in
high-speed liquid chromatography always correspond to the
ascending branch of the H_{eff} versus v curve. If the slope
of the ascending branch is decreasing with increasing
velocities, the highest speed will be achieved with the
maximum velocity, that is, with the maximum available
pressure drop.

If the ascending branch of the H_{eff} versus v curve is linear, the H_{eff}/v versus v presentation converges asymptotically to the slope of the H_{eff} versus v curve. Above the optimal velocity of the eluent, v_{opt}, the speed of analysis increases slightly. In gas chromatography, $v_{opt} \simeq 2v_{min}$, where v_{min} is the velocity at which the ascending branch becomes linear.

It seems to be an empirical fact that in liquid chromatography the H_{eff} versus v curve becomes linear when v >1-3 cm/sec (e.g., Figure 1 in ref. 4 or refs. 8,9,24). In such systems, the optimum linear velocity, v_{opt} is 2-6 cm/sec, more or less independently of the particle size if $d_p > 36$ μ (8).

4. Resolution

With the main assumption that $N_{eff\ 1} = N_{eff\ 2}$ and $(\alpha + 1) = 2\alpha$ (where α is the relative retention), it can be shown that the resolution, R_s, is

$$R_s = \frac{\alpha - 1}{4\alpha} \sqrt{N_{eff\ 2}} = \frac{\alpha - 1}{4\alpha} \frac{k'_2}{1 + k'_2} \sqrt{N_2}$$

Because of the assumption that $N_{eff\ 1} = N_{eff\ 2}$, the equation cannot be used to calculate the resolution of the unretained and the neighboring peak (because $N_{eff\ 1}$ for the unretained peak is zero by definition).

The consequence of the second assumption is that equation 9.7 is valid only if the two peaks to be resolved are not far away from each other. However, adequate and not maximum resolution is wanted in analytical work. If base-line separation is achieved (for symmetrical peaks, $R_s = 1.5$) by increasing the resolution, only the time of analysis will increase; there will be no gain in the quantitative work.

5. Relative Retention, α

On the basis of equation 9.7, the separation of two compounds in a column with poor efficiency, N_{eff}, is possible if α is high enough. Up to now the choice of stationary phases has not been studied enough to solve many of the separation problems in routine work. At the

moment, in the author's opinion, this is the weakest
point in the development of high-pressure liquid chromatography. Fortunately, this shortcoming can be partially
compensated for by programming the parameters during analysis, that is, with a temperature or flow program, or
particularly by using gradient elution.

D.   NEW COLUMNS FOR LIQUID CHROMATOGRAPHY

1.   Stationary Phases

Solid stationary phases are optimal if they are not
soluble in the eluent, as discussed in Chapter 6. They
must be mechanically stable because of the high pressure
and because of the mechanical deterioration caused by the
mobile liquid phase. Alumina, silica gel, and other sil-
ica-type active solids are used primarily. These highly
active solids are hygroscopic, and their activity depends
on their water content. Therefore, the water content of
the solid and that of the liquid mobile phase (and, consequently, the temperature) have to be controlled closely.

If the "liquid" stationary phase is <u>chemically bound</u>
on the support, the stationary phase also must be considered a solid. The advantageous properties of such phases
are described in Chapter 5, and will be discussed later
in this chapter. Often, however, the preparation of such
stationary phases is not simple.

2.   Different Column Types

Most liquid chromatographic analyses are made with
regular packed columns, where $d_c/d_p > 10$. In irregular
packed columns (<u>13</u>-<u>15</u>) the ratio of column diameter to
particle size is smaller than 5. The main advantage of
the irregular over the regular column packing is that,
for a given velocity and particle size, the H or H_{eff}
value remains constant, but the permeability is improved
up to a factor of 10 in irregular packed columns. Some
advantages of these columns in liquid chromatography have
been demonstrated (<u>16</u>).

3. Porous Layer Beads (PLB)

Most of the supports used in chromatography are porous. The advantage of a support with an impenetrable, hard core coated with a thin layer of porous solid (PLB) is the high speed of mass transfer in the stationary phase. This has been pointed out theoretically (16,17) and demonstrated experimentally in gas (18,19,20) and in liquid (21,22, 23,30) chromatography and is discussed in Chapter 1.

For the same linear velocity, v, only about 50% of the flow rate, F, required for porous supports is needed for PLB (with the same inner diameter of both columns), as shown in equations 9.1 and 9.2. The maximum flow rate available from a pump decreases with increasing pressure. Therefore, not only the pressure drop but also the flow rate is a limiting parameter in high-pressure liquid chromatography.

The sample sizes used with PLB columns are small in comparison with those for columns packed with porous supports, because of the small amount of stationary phase present. For the same reason, the capacity ratios, k', of PLB columns are usually small. This is disadvantageous if a complex sample must be resolved, because of the limited peak capacity. At velocities over 1 cm/sec, k' can decrease with increasing velocity if the stationary phase is the porous active solid itself (23). Because of the high efficiency of the PLB, however, this type of column packing is usually advantageous.

Care must be taken if the porous layer beads are coated with liquid stationary phases (24). The PLB packings used in high-speed liquid chromatography have a specific surface area of ca. 1-14 m^2/g of support. However, the specific surface area of the porous layer itself is greater than 100 m^2/g, usually about 300 m^2/g. Solids with such high specific surface areas are active. Coating an active solid with a liquid stationary phase alters the capacity ratio, k', and the relative retention, α, of many substances, causing a variation in the solid to liquid ratio (25,26). Thus, when using liquid stationary phases, one should saturate the liquid mobile phase with

the stationary liquid. If the concentration of the liquid
stationary phase on the solid is smaller than the equilibrium value, stationary liquid phase will be adsorbed
by the PLB until the equilibrium value is reached. During this period, the k' and α values change continuously
(24). On the other hand, if the amount of liquid stationary phase is higher than this equilibrium value, the relative retentions change when the solid to liquid ratio is
varied.

Excellent separations can be achieved with porous layer beads if the stationary phase is chemically bound on
the surface of PLB, as described in Chapter 5.

4. "Brush-Type" PLB

"Brush"-like stationary phases are prepared by the
esterification of silica surfaces with primary alcohols.
The advantageous properties of "brushes" on porous glass
in gas and liquid chromatography have been described (27).
The big disadvantage of these brushes, however, is that
they are not stable to hydrolysis. Consequently, extreme
caution is necessary if the eluent contains proton-donor
liquids, as, for example, hexane with 10% isopropanol
(28). The use of water as the mobile phase results in
the rapid decomposition of this stationary phase; the
"bristles" are "shaved."

In Figure 9.1 the separation of some amines with brushes is shown (29). The stationary phase is a PLB esterified with polyethylene glycol 400. The sample size is
not small (60 μg), and the eluent is 10% isopropanol in
hexane.

It is typical of chromatographic analysis with small
k' values that the holdup time is great, measured on the
separation time, as shown in the figure. Although the
time of analysis itself is short, the linear velocity of
the eluent is only 0.27 cm/sec, usually low in high-speed
liquid chromatography. Some characteristic data for this
separation are given in Table 9.1. If two compounds are
not completely resolved (as B and C in this chromatogram),
the calculated efficiency may appear better (or worse)
than it actually is. As usual with small k' values,

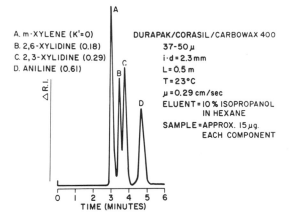

Figure 9.1. Separation of amines on porous layer brush at room temperature (29).

Table 9.1

Characteristic Data for the Separation in Figure 9.1

Parameter	A	B	C	D
k'	-	0.16	0.27	0.55
N	1370	2770	1800	1130
N_{eff}	-	54	80	141
H, mm	0.37	0.18	0.28	0.44
H_{eff}, mm	-	9.3	6.3	3.5
N_{eff}/t, sec^{-1}	-	0.25	0.34	0.38

excellent n values are obtained, and one could argue that the H for compound B is only 3.5 times greater than the particle size (the H/d_p presentation is meaningless indeed, as pointed out earlier). On the contrary, a value for N_{eff} of only 54 is determined from the chromatogram

and the resolution is proportional to $\sqrt{N_{eff}}$. In Table
9.1 it is shown, typically, that N decreases with small
k' values, but N_{eff} increases with increased capacity
ratio. Although the efficiency for compound D is only
N_{eff} = 141, base-line separation is achieved for C and D
because of their high relative retention, α = 2.06. The
relative retention of compounds B and C is smaller, but
high enough for good separation (α = 1.64).

In Figure 9.1 it is demonstrated again that low efficiency, N_{eff}, and high relative retention can result in
speedy and excellent separations. Note the small pressure
drop (50 psi) on the column at v = 0.27 cm/sec. In practice, one can easily elongate the column up to 10 meters
for other separation problems; or, if desired, a combination of a longer column and a faster flow rate will result in shorter analysis time.

The great advantage of chemically bound stationary
phases is demonstrated by the separation in Figure 9.1.
With liquid polyethylene glycol as the stationary phase,
and with hexane/isopropanol as the mobile phase, it is
difficult to maintain equilibrium because of the relatively high mutual solubility.

References

1.   G. Taylor, Proc. Roy. Soc. (London), A219, 186
     (1953).
2.   R. Aris, Proc. Roy. Soc. (London), A235, 67 (1956).
3.   M. J. E. Golay, in "Gas Chromatography, 1958," D. H.
     Desty, ed., Butterworths, London, 1958, p. 36.
4.   L. R. Snyder, J. Chromatog. Sci., 7, 352 (1969).
5.   R. P. W. Scott, D. W. J. Blackburn, and T. Wilkins,
     in "Advances in Gas Chromatography, 1967," A.
     Zlatkis, ed., Preston Technical Abstracts Co.,
     Evanston, Ill., 1967, p. 160.
6.   L. R. Snyder, Anal. Chem., 39, 698 (1967).
7.   I. Halåsz, A. Kroneisen, H. O. Gerlach, and P.
     Walkling, Z. Anal. Chem., 234, 81 (1968).
8.   A. Kroneisen, Ph.D. Thesis, Universität Frankfurt/
     Main, 1969.

9. I. Halász, H. O. Gerlach, A. Kroneisen, and P. Walkling, Z. Anal. Chem., 234, 97 (1968).
10. J. F. K. Huber, in "Advances in Chromatography, 1969," A. Zlatkis, ed., Preston Technical Abstracts Co., Evanston, Ill., 1969, p. 348.
11. I. Halász and P. Walkling, Ber. Bunsenges. Phys. Chem., 74, 66 (1970).
12. G. Deininger and I. Halász, J. Chromatog. Sci., 8, 499 (1970).
13. I. Halász and E. Heine, Nature, 194, 971 (1962).
14. J. C. Sternberg and R. E. Poulson, Anal. Chem., 36, 1492 (1964).
15. I. Halász and E. Heine, in "Advances in Chromatography," Vol. 4, J. C. Giddings and R. A. Keller, eds., Marcel Dekker, New York, 1967, p. 207.
16. J. H. Purnell, Nature, 184, 20009 (1959).
17. M. J. E. Golay, in "Gas Chromatography, 1960," R. P. W. Scott, ed., Butterworths, London, 1960, p. 139.
18. C. Horvath, Ph.D. Thesis, Universität Frankfurt/ Main, Germany, 1963.
19. I. Halász and C. Horvath, Anal. Chem., 36, 1178 (1964).
20. J. J. Kirkland, Anal. Chem., 37, 1458 (1965).
21. C. G. Horvath, B. A. Preiss, and S. R. Lipsky, Anal. Chem., 39, 1422 (1967).
22. J. J. Kirkland, J. Chromatog. Sci., 7, 7 (1969).
23. I. Halász and P. Walkling, J. Chromatog. Sci., 7, 129 (1969).
24. B. L. Karger, H. Engelhardt, K. Conroe, and I. Halász, in "Gas Chromatography, 1970," in print.
25. C. G. Scott and D. A. Rowell, Nature, 187, 143 (1960).
26. I. Halász and E. E. Wegner, Nature, 189, 570 (1961).
27. I. Halász and I. Sebestian, Angew. Chem., Intern. Ed., 8, 453 (1969).
28. C. G. Scott, in "Advances in Chromatography, 1970," in print.
29. J. N. Little and D. F. Horgan (Waters Associates), unpublished results.

30. B. L. Karger, K. Conroe, and H. Engelhardt, J. Chromatog. Sci., 8, 242 (1970).

PART THREE
APPLICATIONS OF LIQUID CHROMATOGRAPHY

CHAPTER 10

A Comparison of Separation Mechanisms for
Liquid Chromatography Applications

Karl J. Bombaugh

A. INTRODUCTION

Up to this part of the book we have been concerned
primarily with the theory and practice of high-performance liquid chromatography. We have considered the various techniques as they apply to the four major separation
mechanisms, that is, to adsorption, partition, exclusion,
and ion-exchange. We have also considered hardware requirements, devoting particular attention to detectors.
This book would not be complete, however, particularly
in the eyes of the novice, without reducing all of this
technology to simple practice, directing attention to
which technique to use and how to use it for a given application.

The novice may well ask such questions as, "Which
mechanism should I use, and why?" "If one technique is
fast, why use another which is slower?" "Is one technique really slower than another?" This chapter will
attempt to shed light on the matter by showing separations of a given material by each of several mechanisms
to illustrate what is accomplished by a particular mechanism. Some advantages of each mechanism will also be
discussed. We will consider how the respective mechanisms relate to each other in performance and, finally,
how they may be used to advantage in conjunction with one
another.

B. LINEAR ELUTION VERSUS SOLVENT-PROGRAMMED
LIQUID-SOLID CHROMATOGRAPHY

Polyglycols are commonly used as nonionic surfactants
and as lubricants. These materials are complex mixtures
which range between 200 and 2000 in molecular weight and
differ from one another by perhaps 40 molecular weight
units. The chromatographic illustrations shown here were
not prepared specifically for the comparison of these

two techniques, but were selected from other published
works and are assembled here to illustrate the principles.

Figure 10.1 shows a separation of Tergitol® on a 0.303-
in.-i.d. column of a porous adsorbent, Porasil® 60. The

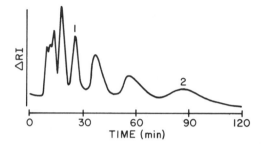

Figure 10.1. Liquid-solid chromatogram of Tergi-
tol® TMN. Each peak represents an increase of a
single ethylene oxide unit on the alkylphenyl base.
Conditions - column: 1' x 0.303" i.d., packed with
37-62 μ Porasil® 60; carrier: methyl ethyl ketone
containing 2.5% water; flow rate: 1 ml/min. Ref.:
K. J. Bombaugh et al., Res./Develop., 9, 28 (1968).

general elution problem (1) is clearly evident.* The
first three peaks are unresolved; the next four peaks are
well resolved, while the last peak shown is as wide as
the three preceding resolved peaks. The chromatogram
provides no assurance that the last peak shown is, in
fact, the last, and no answer to the question of whether
or not additional oligomers are held up in the column to
elute later as wider peaks. Figure 10.2 shows the same

---

* The term was applied by Snyder to describe a problem
common to all types of sorption chromatography, that is,
the fact that no solvent of a given polarity is optimum
for all components in a mixture containing a wide range
of distribution coefficients and therefore eluting over
a wide range of k'. In short, no general eluent exists.

separation made with programmed solvent operation. This separation, also done with a 3/8 in.-i.d. x 1 ft column of Porasil® , has accomplished the separation of all components. Solvent programming has solved the general elution problem. Early peaks are resolved by using a relatively nonpolar solvent mixture, which is gradually increased in polarity to elute the slower-moving components from the column.

A logarithmic gradient may be generated by a system which feeds the solvent pump, such as the Varigrad® (2) solvent-programmer system. The logarithmic-type program produces nearly equal peak widths, analogous to temperature programming in gas chromatography. The use of highly polar carriers at the conclusion of the separation provides reasonable assurance that the last peak is eluted.

It should be stressed that the column must be regenerated before reuse when carrier programming is employed. Regeneration is best accomplished by pumping several volumes of starting solvent through the column to re-establish starting equilibrium. Solvent-programmed operation combined with the use of porous supports in large-diameter columns is ideal for preparative chromatography. High column capacity and large column volume permit heavy sample loads. Solvent programs permit elution of a wide range of capacity ratios (k') over a comparatively short time. Fractions collected across the distribution of k's are available for additional chromatographic treatment, or for examination by ancillary techniques such as infrared and ultraviolet spectrophotometry.

Use of the differential refractometer detector with solvent programming (Figure 10.2) was accomplished by employing dual, parallel columns in conjunction with polar and nonpolar solvents, which are similar in refractive index (3). In this work isopropanol and hexane, which differ in RI by 0.004 RI unit, were found suitable. Isorefractive mixtures can also be used. For example, n-heptane can be added to the n-hexane to adjust the RI of the nonpolar solvent to that of the isopropanol. In this way less burden is placed on the operator to match

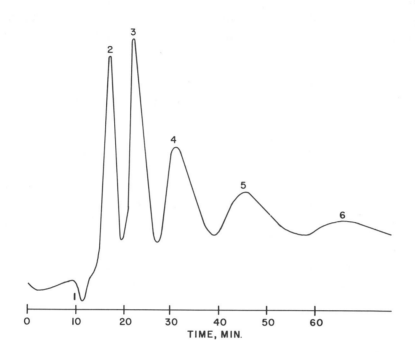

<u>Figure 10.2.</u>   Solvent-programmed separation of
Ucon® 50HB55. <u>Conditions</u> - column: 1' x 0.303"
i.d., packed with 37-62 μ Porasil® 60; solvent
program: 10-70% isopropanol in <u>n</u>-hexane; flow
rate of sample and reference columns: 1 ml/min;
detector: differential refractometer with dual
column - matched flow operation. <u>Ref.</u>: K. J.
Bombaugh et al., J. Chromatog., <u>43</u>, 332 (1969).

the flows of the dual-column system, so that the detector base line remains stable.

Figure 10.3 shows the separation of another surfactant, Triton® X-45, on porous layer beads (Corasil® II).
Here the peaks are resolved in 10 min, and the first four
in less than 5 min. Although the general elution problem
is still evident, it is not as severe with the solid core
packing as with the large-particle porous supports shown
previously.

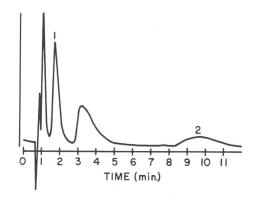

Figure 10.3. Isocratic separation of Triton® X-45
on porous layer adsorbent. Conditions - column:
50 cm x 2 mm i.d. of Corasil® II; carrier: n-hexane containing 10% isopropanol; flow rate: 1.55
ml/min.

A comparison of separation performance by the two packings as related to peaks 1 and 2 in Figures 10.1 and 10.3
is shown in Table 10.1. The porous layer packing affords
a ten-fold increase in effective plates per second over
the large-particle porous packings (see also Chapter 1).
Since column efficiency at high flow rates does not tend
to degrade a severely retained peak on porous layer packings, a wider range of materials can be separated by a
single solvent. For high-speed linear-elution liquid-
solid chromatographic analysis the porous layer beads are
ideal, whereas for high-capacity preparative chromatography porous packings are preferred. However, for optimum resolution, the latter packings should be run at
lower solvent velocities.

C. LIQUID-SOLID VERSUS GEL PERMEATION CHROMATOGRAPHY

A high-resolution GPC separation of Triton® X-45 and
X-100 is shown in Figure 7.4 (Chapter 7). This separation was made on a very long column (160 ft) packed with
500 Å Poragel® and required 60 hr. This separation is

Table 10.1

A Comparison of Separation Performance
with Porous and Porous Layer Packings (4)

Packing	Peak	k'	Time, sec	x 10³ k'/t, sec	N, Plates	H, mm	v, cm/ sec	N/t, Plates/ sec
Porous	1	2	1540	1.3	225	2.22	0.06	0.15
Porous	2	8.9	5080	1.8	112	4.46	0.06	0.02
Porous layer	1	1.8	99	18.3	104	4.80	1.35	1.05
Porous layer	2	14.5	548	29.2	114	4.40	1.35	0.21

slow when compared to high-speed LSC, not because GPC itself is slow, but because the low selectivity (α = 1.05) available required the use of a long column. (For comparison with liquid-liquid chromatography, see Figure 11.2.) This technique has provided both separation and size information. The relationship between Triton® X-45 and X-100 is clearly evident from the elution volume. Since the peak widths are equal across the distribution, the number of resolvable species is evident.

As discussed in Chapter 6, liquid-solid separations can be made by column chromatography in the same time as is required for thin-layer chromatography, but with the added advantage of quantitative response and superior resolution. The chromatogram in Figure 10.4 shows the separation of a polypropylene glycol sample (Ucon® 50HB55) in 8 min, using a porous layer packing. This separation is clearly superior to the TLC separation shown at the bottom of the figure. With the resolution shown in the column separation, quantitation is readily possible. It should be pointed out, however, that, since Ucon® exhibits no ultraviolet chromophore, a detector such as the differential refractometer is mandatory.

D. ADSORPTION VERSUS PARTITION CHROMATOGRAPHY

The separation of ten insecticides shown in Figure 10.5 was accomplished, using a 37-μ porous support, by classical liquid-partition chromatography. Figure 10.6 shows a separation of several of the same insecticides made by liquid-solid chromatography, using a porous layer adsorbent column. Some important principles can be learned from a comparison of these chromatograms. Perhaps the first lesson is that the high speed is not determined by the chromatograph, but rather by the column, since the same instrument was used for both separations. The five components separated in 50 min on the porous packing are separated in 8 min with the porous layer packing. Lindane (peak 9), which elutes in 40 min from the porous packing, elutes in less than 3 min (peak 5) from the porous layer packing.

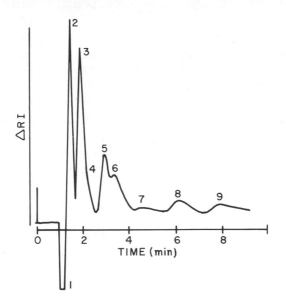

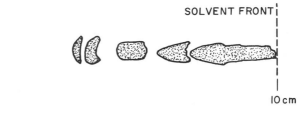

Figure 10.4. Comparison of high-performance column chromatography with thin-layer chromatography for separating polypropylene glycol (Ucon® 50HB55). Top - column: 50 cm of Corasil® II; carrier: n-hexane containing 5% isopropanol; flow rate: 0.9 ml/min; sample: 0.8 mg in 20 µl solvent; complete separation: 9 min. Bottom - (reproduction of a thin-layer plate) - developing solvent: n-hexane containing 20% ethanol; spot developer: iodine vapor; time: 40 min.

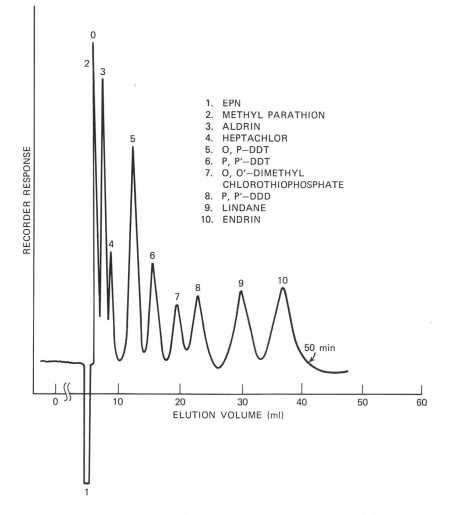

Figure 10.5. Liquid-liquid chromatogram of pesticides using porous support. Conditions - column:
4' x 0.093" i.d., with 10% β,β-oxydipropionitrile
on 37 μ of Porasil® 60; carrier: isooctane saturated with β,β'-ODPN; flow rate: 0.85 ml/min. Ref.:
D. L. Horgan et al., Waters Associates, TR #19531.

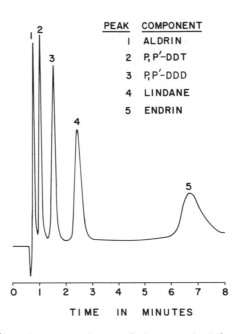

PEAK	COMPONENT
1	ALDRIN
2	P,P'-DDT
3	P,P'-DDD
4	LINDANE
5	ENDRIN

TIME   IN   MINUTES

<u>Figure 10.6</u>. Separation of insecticides on porous
layer adsorbent. <u>Conditions</u> - column: 50 cm x 2.3
mm i.d., of Corasil® II; carrier: <u>n</u>-hexane; flow
rate: 1.5 ml/min. <u>Ref</u>.: Bombaugh et al., J. Chromatog. Sci., <u>8</u>, 657 (1970).

The separation in Figure 10.6 is by adsorption, with
no stationary liquid phase being used. No presaturator
column was required, and there is no concern for column
bleed. In addition, temperature control is less vital,
since adsorption coefficients are not as sensitive to
temperature as partition coefficients. In partition
chromatography elution volume is a function of stationary
liquid load, which varies with temperature since the solubility of the stationary phase in the moving phase varies
with temperature. Adsorption coefficients are also sensitive to temperature variation, but the effect on retention volume is usually less.

To this extent, then, when resolution is permitted by
adsorption, the latter is the preferred mechanism of separation because it is simpler. One should not infer,

however, that the LSC separation is inherently faster than the LLC separation. The increase in speed is afforded by the porous layer packing, rather than by the mechanism of retardation. A polar modifier normally required with porous adsorbents (see Chapter 6) was not needed for this separation using the porous layer adsorbent.

In spite of the restrictions of phase stability and limitations in the availability of miscible pairs, liquid-partition chromatography affords separations not always possible by other techniques. Properly chosen solvent pairs can afford unique separations, and in such cases liquid-partition chromatography is the preferred technique. Figure 10.7 shows the liquid-liquid chromatogram of an extract from a perch. Pure lindane was added to the extract as a marker, and the chromatograms were superimposed.

E. ADSORPTION CHROMATOGRAPHY WITH POROUS LAYER ADSORBENTS

Separation by adsorption is both simple and fast, particularly with porous layer or small-particle adsorbents. The separations shown in Figure 10.8 were made by adsorption on the porous layer packing, Corasil® II, using chloroform as a solvent, the adsorbent being activated at 110°C for 3 hr before packing the column. These chromatograms illustrate the effect of carrier flow rate on the resolution of these vitamin constituents. By tripling column length, vitamin D_2 is partially separated from vitamin A in 15 min.

Figure 10.9 shows the separation of a mixture of barbiturates, also by adsorption on Corasil® II, using chloroform as a solvent. The outputs of two detectors, a UV photometer and a refractometer, are superimposed for illustrative purposes.

When separation is not possible with silica, other adsorbents, such as alumina, charcoal, and reversed-phase adsorbents [e.g., "brushes" developed by Halász (5) or silicone derivatives developed by Kirkland (6)] are useful. Brushes are organic derivatives of silica in which alcohols are attached to the support by a silicate ester

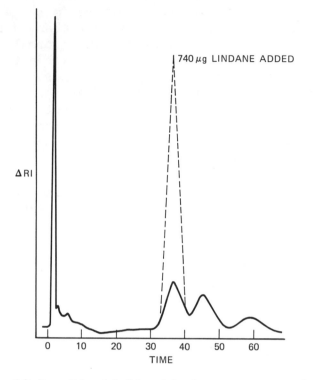

Figure 10.7. Liquid-liquid chromatogram of fish
extract enriched with lindane. Conditions - column: 3' x 0.093" i.d., packed with 10% β,β'-oxy-
dipropionitrile on 37 μ of Porasil® 60; carrier:
isooctane; sample: 20 μl of extract representing
1.8 g of fish. Ref.: Horgan et al., Waters
Associates TR #19531.

bond. Highly stable silicon-to-carbon bonds are characteristic of the silicone derivatives.

F. ADSORPTION CHROMATOGRAPHY WITH SPECIALIZED PACKINGS

1. Chemically Bonded Supports ("Brushes")

Unique separations may be performed with specialized
adsorbents having chemically bonded organic surfaces.

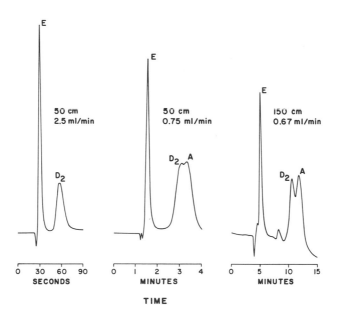

Figure 10.8. Effect of flow rate on separation of fat-soluble vitamins, using porous layer adsorbents. Conditions - column: Corasil® II - length as specified; carrier: chloroform (flow as shown). Ref.: Bombaugh et al., J. Chromatog. Sci., **8**, 657 (1970).

Figure 10.10 shows the separation of endrin from endrin ketone on Durapak®/Carbowax® 400. Attempts to separate these compounds with conventional adsorbents were unsuccessful. The endrin ketone is not soluble in solvents such as isooctane, which meets the immiscibility requirement to be used with β,β'-oxydipropionitrile. The insecticide pair is soluble in xylene, but xylene is miscible with the known stationary liquid phases. By using a chemically bonded packing, Durapak®/OPN-Porasil® 400, in which the oxypropionitrile is attached to the silica by an ester bond, this separation may be carried out with xylene as a carrier.

Figure 10.11 shows the separation of positional isomers with a Durapak®/Carbowax® column. Although the

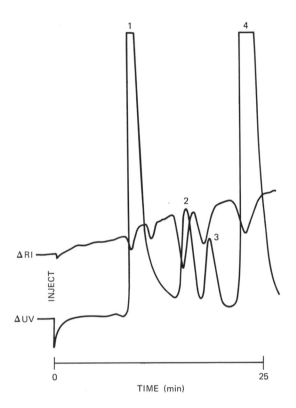

Figure 10.9. Separation of barbiturate mixture on
porous layer adsorbents, using tandem detectors.
Conditions - column: 6' x 0.093" of Corasil® II;
carrier: chloroform; flow rate: 0.5 ml/min.
Components - 1: unknown; 2: pentobarbital; 3: seco-
barbital; 4: phenobarbital.

Porasil® support used with Durapak® materials does not
afford the efficiency of the solid core packings, the
high selectivity permits a separation of these mixtures
in less than 6 min. The chief limitation of the Durapak®
is that the ester bond is extremely labile and undergoes
hydrolysis when exposed to water, or undergoes alcohol
exchange when exposed to lower alcohols. Although
Durapaks® may be used with solvents containing small

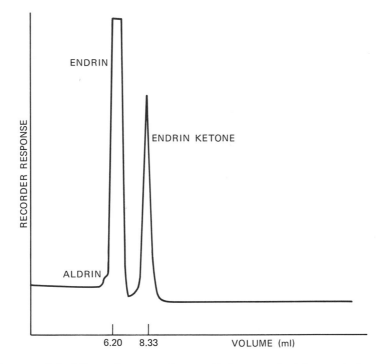

Figure 10.10. Separation of insecticides with chemically bonded oxypropionitrile packing. <u>Conditions</u> - column: 6' x 0.093" i.d. of Durapak®/ OPN-Porasil® 400; carrier: xylene; flow rate: 0.3 ml/min; sample: 3 mg. <u>Ref</u>.: W. A. Dark, Washington University Short Course in LC, May 1970.

quantities of alcohol (10-20%), they must be used iso-cratically. For example, once used at a 10% ethanol level, Durapak® cannot be used at 5% without some tailing for retained components.

2. Chemically Bonded Porous Layer Beads

It is desirable to gain both the improved mass transfer in the stationary moving phase afforded by porous layer beads and the selectivity exhibited by the chemically bonded organic surfaces. This combination is

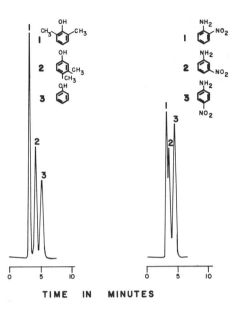

TIME IN MINUTES

<u>Figure 10.11</u>. Liquid chromatogram of positional
isomers on Carbowax® 400 chemically bonded to a
porous support. <u>Conditions</u> - column: 3' x 0.093"
of Durapak®/Carbowax® 400; carrier: chloroform;
flow rate: 1.05 ml/min. <u>Ref</u>.: Bombaugh et al.,
J. Chromatog. Sci., <u>8</u>, 657 (1970).

accomplished by chemically bonding the organic phase to
the porous layer support (<u>6,7</u>). The net result is illustrated in Figure 10.12, which shows a separation carried
out with Carbowax® 400 chemically bonded to Corasil®.
Four components are separated in less than 2 min, indicating high column efficiency at high carrier flow rate
with the porous layer "brush" column.

   3.   <u>Porous Polymers as Reversed-Phase Sorbents</u>

   Porous polymers, commonly used for size separation by
exclusion chromatography, can be used to advantage for
reversed-phase sorption chromatography, as shown in Figure 10.13. Poragel® P N, a cross-linked porous polymer,

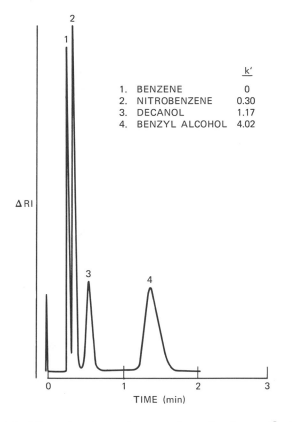

Figure 10.12. Separation using Carbowax® 400 chemically bonded to a porous layer base. Conditions - column: 50 cm x 2.3 mm i.d. of Carbowax® 400-Durapak®/Corasil®; carrier: n-hexane; flow rate: 3.1 ml/min. Ref.: Little et al., J. Chromatog. Sci., 8, 625 (1970).

is used with a methanol-water mixture (polar solvent) to separate steroids. Although the efficiency for this column is comparatively low (N_{eff} = 100 for cyasterone), resolution is afforded by the high selectivity. The response curves of both the UV and the RI detectors are shown.

These nonbleeding packings are ideal for quantitative liquid chromatography since the columns are highly stable.

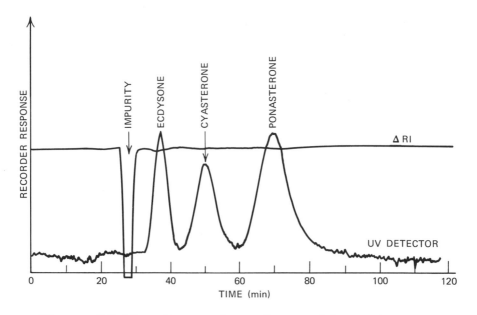

<u>Figure 10.13</u>.   Separation of steroids, using por-
ous polymers as reversed-phase sorbents.   <u>Condi-
tions</u> - column: 3' x 0.303" i.d. of Poragel® P N;
carrier: methanol containing 30% water; flow rate:
1 ml/min.   <u>Ref</u>.:   W. A. Dark, Washington University Short Course in LC, May 1970.

The uniform retention times afforded by these stable columns also afford reproducible peak heights.   The porous
polymers are subject, however, to the same advantages (in
capacity) and disadvantages (in mass transfer) described
previously for other porous packings.   The high resistance
to mass transfer is evident in the decrease in N from 370
to 104 accompanying the increase in k' from 0.5 to 1.7.
This suggests that, although porous polymers can provide
excellent selectivity, they afford limited peak capacity,
particularly at high solvent velocity.

G.   BASIC ADSORBENTS

Some forms of alumina are basic materials and, therefore, offer a retention characteristic different from the
acidic silica surface.   Therefore, alumina deserves

consideration as a packing in high-performance liquid
chromatography. Figure 10.14 shows a separation of an
"air pollution" sample on an alumina column. The Turner
Fluorometer was used in conjunction with the UV detector
in a Waters Associates ALC-301 instrument. The flow cell,
fabricated especially for this experiment, was not optimum. However, in spite of the excessive peak spreading
in the flow cell, the sensitivity of the fluorometer operated at optimum wavelength far exceeds that of the highly
sensitive UV detector operated at 254 mμ.

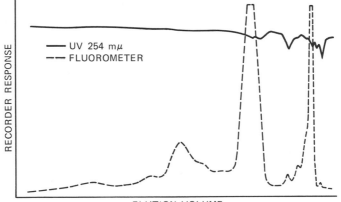

Figure 10.14. Chromatogram of air pollutants, using UV and fluorescence detection. Conditions -
column: 6' x 0.093" i.d. of alumina; carrier: n-
hexane; flow rate: 0.25 ml/min. Ref.: W. A. Dark,
Washington University Short Course in LC, May 1970.

H. EXCLUSION CHROMATOGRAPHY

In this section, final consideration is given to the
applications of gel permeation chromatography. The GPC
technique is of great value because elution volume is
related to molecular size and results are predictable.
Once the calibration curve has been developed, the elution volume and, therefore, the elution times of the
first and the last peak are known. In addition, the size

function of each molecule is known as well. For this
reason, GPC provides a unique capability in that it can
separate and provide size information simultaneously.
   Since most molecules change size during a chemical
reaction, GPC is an ideal way to follow either synthesis
or degradation. During a kinetic study, if a sample withdrawn from a reactor shows either a peak shift to a lower
elution volume, or a new peak at a lower elution volume,
it is clear that a higher molecular weight substance is
formed. This judgment is valid without the need for additional separation or identification. Furthermore, such
results can be obtained with comparatively low resolution
at very high speed. By relating elution volume to molecular weight or size, reliable judgments can be made which
greatly simplify qualitative analytical problems. The
combination of preparative liquid chromatography with adsorption or mass spectrometry is a powerful tool for the
identification of mixtures. However, when judgments concerning the molecular size of the eluents can be made
directly from the chromatogram without the need of collecting fractions, as is common with GPC, considerable
time and labor can be saved in problem-solving.

1. Gel Permeation Chromatography as a "Molecular Size Spectrometer"

   The series of chromatograms in Figure 10.15 represents
a series of groups of fatty acid esters, hexatol anhy-
drides, produced under the names of Span® and Tween®.
With no prior knowledge of the composition of the materials, it was possible to learn a great deal from these
GPC chromatograms. The parent peak, eluting at 135 ml,
is immediately known to represent a larger molecule than
those eluting at 190 ml. When the material is compared
to the triglycerides shown at the top of the figure, it
is evident that the material is similar in size to tristearin. It is clear that the trioleate contains monooleate, and that the monooleate contains some trioleate.
The monolaurate is easily distinguishable from the palmi-
tate and other large molecules.

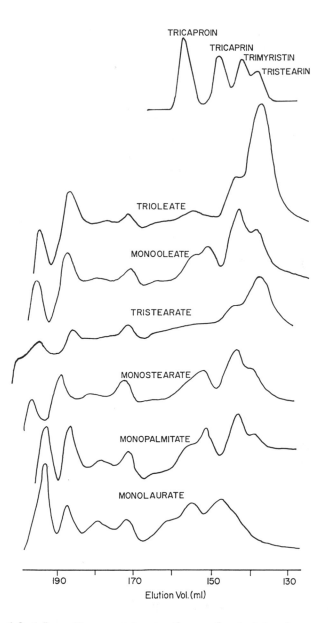

Figure 10.15. Separation of surfactants by exclusion chromatography. Conditions - column: 20' x 0.303" i.d. of 500 Å Poragel®; carrier: tetrahydrofuran; flow rate: 1.1 ml/min; sample: fatty acid

Figure 10.15 (cont.). esters of hexatol anhydride.
Ref.: Bombaugh, et al., Z. Anal. Chem., <u>236</u>, 443
(1968).

The smaller molecules eluting beyond 170 ml may be
either reaction products or additives. However, before
making a preparative separation and collecting fractions,
it may be adequate to obtain an infrared spectrum of the
total sample and combine this with process knowledge to
make the assignments. This technique has been used by
Spell (<u>8</u>), who combined molecular size and IR spectral
information to provide qualitative analysis.

2. Gel Permeation Chromatography as a
<u>Distillation Analyzer</u>

The chromatogram of crude oil in Figure 10.16, before
and after distillation, shows the potentiality of GPC as

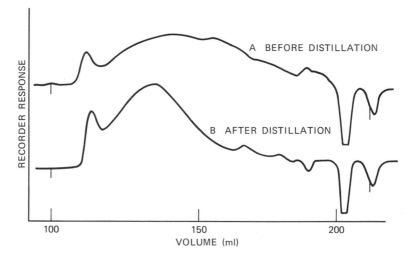

Figure 10.16. Exclusion chromatogram of crude oil.
Conditions - column: 20' x 0.303" i.d. of 500 Å
Poragel®; carrier: tetrahydrofuran; flow rate: 1
ml/min; sample: 2.5 mg. Ref.: Bombaugh, et al.,
Separation Sci., <u>3</u> (4), 375 (1968).

a distillation analyzer for crude oils to replace or sup-
plement distillation and gas chromatography. The method

usable at room temperature, is nondestructive of any part of the sample and is highly reproducible. Moreover, it is easily accomplished with simple equipment.

3. Miscellaneous Gel Permeation Analysis

Reactive Compounds. Size separation of a hydroperoxide catalyst mixture is shown in Figure 10.17. These

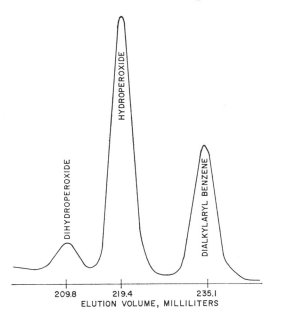

Figure 10.17. Separation of a catalyst mixture by exclusion chromatography. Conditions - column: 15' x 0.303" i.d. of 500 Å Poragel® plus 12' of 100 Å Poragel® ; carrier: tetrahydrofuran; flow rate: 0.9 ml/min; sample: 2 µl. Ref.: Bombaugh et al., J. Chromatog. Sci., 7, 42 (1969).

highly reactive materials are used as initiators in polymerization reactions. The gentle action of gel permeation assures separation of this and other sensitive compounds without conversion or rearrangement, such as is known to occur on active surfaces or under strong "bonding" influences. Such analytical separations are now in use industrially.

Broad Distribution Mixtures. Gel permeation chromatography is widely used to analyze polymer additives and
polymer extracts. Figure 10.18 shows a mixture of materials used as a plasticizer in polyvinyl chloride. Figure 10.19 shows a mixture of polyethylene extract, including an antioxidant and a slip agent. Some of these

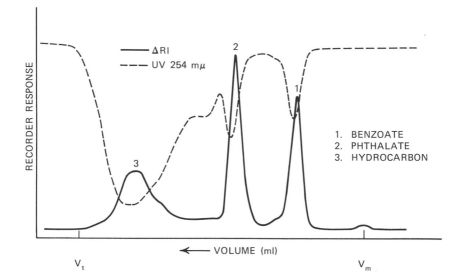

Figure 10.18. Exclusion chromatogram of plastici-
zers extracted from polymer. Conditions - column:
30' x 0.303" i.d. of Poragel® (3' of 500 Å, 12' of
100 Å, 15' of 60 Å); carrier: tetrahydrofuran;
flow rate: 1 ml/min. Exclusion volume (V_m) = 208
ml; total permeation volume (V_t) = 375 ml.

materials may be detected by gas chromatography; others
cannot. However, the real need is to separate and determine the materials in the polymer extract. To do this by
gas chromatography requires a preliminary separation and
sometimes the preparation of derivatives. With GPC the
entire mixture is accomplished in 2 hr of elapsed time
with no prior separation required. The gel permeation
chromatogram provides a profile of the entire extract,
which includes the low-molecular-weight polymer and the

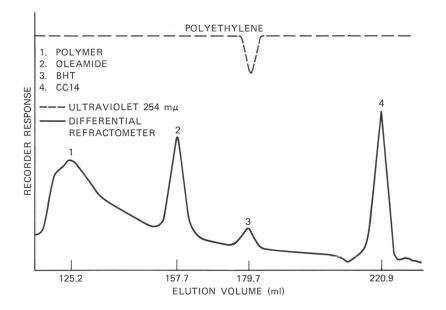

Figure 10.19. Exclusion chromatogram of extract from polyethylene. <u>Conditions</u> - column: 21' x 0.303" i.d. of 60 Å Poragel®; carrier: tetrahydrofuran; flow rate: 1 ml/min.

additives. This analysis provides a direct approach to the material extractable from the polyethylene film.

Motor oils may be characterized by GPC as shown in Figure 10.20. With no prior knowledge of motor oil technology, considerable information can be obtained from the chromatograms. Company F (line C) uses a narrow molecular weight distribution of oils to make a multiviscosity oil and employs a high-molecular-weight viscosity builder (at V_R = 100 ml) plus an additive, which was found to disappear from the oil after use (line B). Gel permeation chromatography can also be used in a kinetic study; for example, GPC of crankcase samples after each 100 miles can establish the rate of consumption of this additive. Company L (line A) uses material of much higher molecular weight and a much wider distribution in their multiviscosity oil. The shoulder of the high-molecular-weight end may also be a viscosity builder.

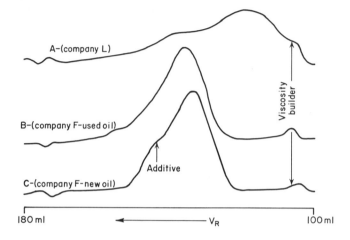

Figure 10.20. Exclusion chromatograms of multiviscosity motor oils. $_{\circ}$Conditions - column: 18' x 0.303" i.d. of 100 Å Poragel®; carrier: tetrahydrofuran; flow rate: 1 ml/min; sample: 2.5 mg. Ref.: Redrawn from Figure 4, Bombaugh et al., Res./Develop., 9, 28 (1968).

Figure 10.21 shows a chromatogram of mineral oil as obtained with two detectors in series. The UV detector, which is sensitive to the aromatic material, shows a distribution completely different from that of the RI detector. These results suggest that the aromatics are lower in molecular weight than the naphthenes and paraffins. Since the RI instrument is often considered as a general detector, the RI response curve may erroneously be regarded as the mass distribution of the sample. Caution in interpretation is needed, since the RI response is a function of the difference in RI between the solvent and the solute. Since the RI of aromatics is different from that of paraffins and naphthenes of the same molecular weight, a distorted RI response curve can be produced.

Albaugh (9) observed that an apparent bimodal distribution of crude oil was in reality a log-normal

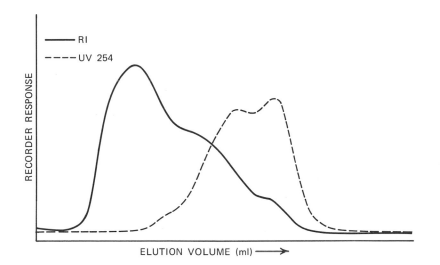

Figure 10.21. Exclusion chromatogram of mineral oil, comparing responses of a differential refractometer and a UV detector. Conditions - column: 21' x 0.303" i.d. of 60 Å Poragel®; carrier: tetrahydrofuran; flow rate: 1 ml/min; sample: 20 mg. $V_t \cong 220$ ml; $V_m \cong 120$ ml.

distribution when determined by the mass of solute of equal fractions, as illustrated schematically in Figure 10.22. When the sensitivity difference is known, the detector can be calibrated and the curve corrected. Furthermore, the UV detector can be used to distinguish classes, providing additional information regarding the distribution.

Aqueous GPC with Rigid Gels. At the present time high-speed GPC with aqueous solvents can be carried out with rigid siliceous packing, such as Porasil®, deactivated Porasil®, and Corning CPG® (controlled-porosity glass). Figure 10.23 shows a separation of sodium polysilicate. No other packing was found from which this material will elute. In this system the components in the mixture are eluted with symmetrical peaks and with K_0 values less

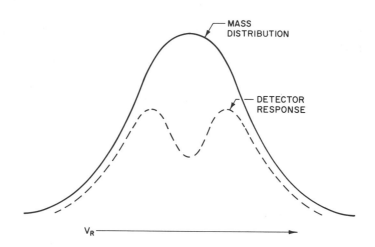

<u>Figure 10.22</u>. Symbolic representation of a ΔRI response curve compared to a mass distribution curve for a mixture of aliphatic and aromatic hydrocarbons, with benzene as carrier. <u>Ref</u>.: Redrawn from Figure 7, Albaugh et al., Preprints, Div. Petroleum Chem., Am. Chem. Soc., <u>15</u>, (2), A225 (February 1970).

than 1, showing no adsorption. The carrier used was water with no adsorption suppressor.

Figure 10.24 shows an aqueous GPC chromatogram of lignin sulfonate separated on deactivated Porasil®. Previous work with lignin sulfonate showed considerable evidence of adsorption. However, with deactivated Porasil®, elution apparently occurs by size only. A chromatogram of polypropionic acid, as obtained with Corning's CPG®, is shown in Figure 10.25. The uniform pore size of Corning CPG® provides high resolution, while the glass base elutes the polar solute without need for an adsorption suppressor.

I. SUMMARY

The applications described above were selected to illustrate the capability of each technique. No technique is superior in all areas. The responsibility to select the technique which provides optimal advantage

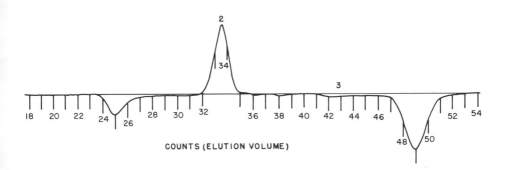

Figure 10.23. Aqueous GPC chromatogram of polysodium silicate on porous silica. Conditions -
column: 16' x 0.303" i.d. of deactivated Porasil®
60; carrier: 0.1 N nitric acid; flow rate: 1 ml/
min; sample: 20 mg.

V_R Counts*	Degree of polymerization
25.0	~ 1500
33.6	~ 100
42.0	~ 30
49.0	~ 80

*Count numbers represent 5 ml of retention volume.
 Each count of elution volume = 5 ml.
Ref.: Bombaugh, Dark, and Little, Proceedings 7th
Intern. Seminar on GPC, Monaco, October 1969.

for the problem at hand rests with the chromatographer.
The hazard facing the practitioner is the tendency to use
a familiar technique for all or most applications and to
neglect the others.

For complex separation problems, maximum advantage is
usually gained by utilizing the techniques in combinations wherein they complement one another. Figure 10.26
expresses the interrelationships between the various
techniques graphically. It is included here to serve as

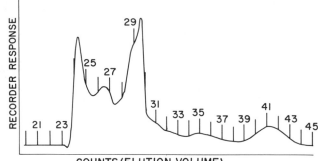

Figure 10.24. Aqueous GPC separation of lignin sulfonates on porous silica. Conditions - Column: 16' x 0.303" i.d. of deactivated Porasil® (equal lengths of Porasil® 1000, 400, 250, 60); carrier: water; flow rate: 1 ml/min; sample: 5 mg. Ref.: Bombaugh, Dark, and Little, Proceedings 7th Intern. Seminar on GPC, Monaco, October 1969.

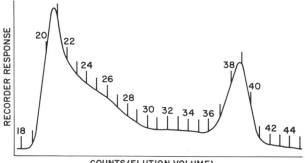

Figure 10.25. Aqueous GPC chromatogram of polypropionic acid on controlled-porosity glass. Conditions - column: 16' x 0.303" i.d. of Corning CPG® (CPG 10-1250, 10-700, 10-240); carrier: 0.05M NaCl at 1 ml/min; sample: 5 mg. Ref.: Bombaugh, Dark, and Little, Proceedings 7th Intern. Seminar on GPC, Monaco, October 1967.

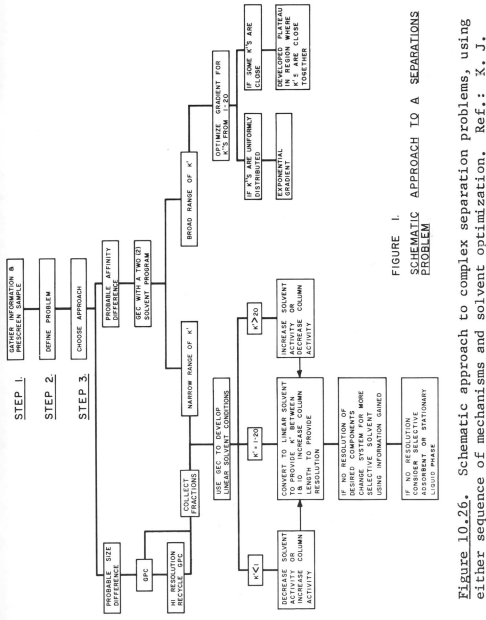

Figure 10.26. Schematic approach to complex separation problems, using either sequence of mechanisms and solvent optimization. Ref.: K. J. Bombaugh, Am. Lab., 7, 43 (1969).

373

a stimulous to the reader, who may refine the chart as
he gains experience in developing his own programs.

References

1. L. R. Snyder, "Principles of Adsorption Chromatog-
    raphy," Marcel Dekker, New York, 1968, pp. 9-38.
2. E. A. Peterson and H. A. Sober, Anal. Chem., 31, 857
    (1959).
3. K. J. Bombaugh, R. N. King, and A. J. Cohen, J.
    Chromatog., 43, 332 (1969).
4. K. J. Bombaugh, "A Consideration of Factors Affecting
    Analysis Time in Liquid Chromatography," Chromatog.
    Notes, 1, (1), 13 (1970).
5. I. Halász, and I. Sebestian, Angew. Chem. Intern.
    Ed., 8, 453 (1969).
6. J. J. Kirkland and J. J. DeStefano, J. Chromatog.
    Sci., 8, 309 (1970).
7. J. N. Little, D. L. Horgan, and K. J. Bombaugh, "A
    Comparison of Conventionally Coated and Chemically
    Bonded Stationary Phases in Liquid Chromatography,"
    presented at 160th Meeting of the American Chemical
    Society, Chicago, Sept. 13-18, 1970.
8. J. L. Spell, Proceedings 4th Intern. Seminar on Gel
    Permeation Chromatography, Miami, 1967, p. 180,
    published by Waters Associates, Framingham, Mass.
9. E. W. Albaugh, P. C. Talarico, B. E. Davis, and R. A.
    Wirkkala, Preprints, Div. Petroleum Chem., Am. Chem.
    Soc., 15, (2), A225 (1970).

CHAPTER 11

Applications of High-Speed Liquid Chromatography Using Controlled Surface Porosity Support

John A. Schmit

A. INTRODUCTION

A number of recent developments have made liquid chromatography attractive for the solution of a variety of analytical problems. In the past, this technique has generally been limited by the column efficiencies obtainable. The limited efficiency resulted in long analytical times to achieve desired separations. The long analytical times, in turn, usually meant dilution of the sample by the mobile phase, requiring large sample sizes for adequate detection. These disadvantages, as well as serious limitations in instrumentation, have kept liquid chromatography applications from keeping pace with the use of gas chromatography. However, a number of recent developments in the field have rapidly expanded the routine application of liquid chromatography to a wide variety of analytical problems.

Notable among these advances has been the development of superficially porous (or porous layer) support materials. These supports have provided mechanically stable, highly efficient columns which produce high-speed separations. Concurrently, a great deal of exploration of stationary phases has also taken place. In many cases, the new stationary phases introduced, along with the high-performance support materials, have opened additional areas of application for liquid chromatography. The types of solid-core support materials available were described in Chapter 5. A comparison of two of these, Corasil®* and Zipax®**, as support materials for liquid-liquid partition chromatography was reported by Majors (1).

---

* Waters Associates trademark.
**E. I. du Pont de Nemours & Company trademark.

The separations discussed in this chapter represent only a few of the possible applications of liquid chromatography in a number of fields, but are indicative of the speed and resolution attainable with modern instrumentation and column technology. The particular chromatographic system selected likewise does not represent the only solution to the separation problem. In most of the examples cited, the mixtures were chromatographed to establish the feasibility of the separation rather than to optimize the analysis.

All of the separations reported here were performed in the Du Pont Instruments Applications Laboratory, using Zipax® chromatographic support described by Kirkland (2,3), and instrumentation previously described (4). To gain insight into how the separations were achieved, a brief description of each of the stationary phases used and the important mobile-phase variables for each will be given.

The liquid-liquid partition systems described in this chapter consist of a stationary phase coated at 1-2% by weight on the superficially porous solid support, a diatomaceous earth precolumn coated with 30% by weight of the stationary phase, and a mobile phase saturated with the stationary phase. The most common mobile phases are cyclopentane, hexane, heptane, and isooctane. The retention time of the sample is adjusted by modifying the mobile phase with small amounts of a more polar solvent, usually up to 10% by volume. Ideally, the modifier is a good solvent for the sample and is miscible with the mobile phase.

The most useful liquid-liquid systems will be briefly described and compared in the following sections. The efficiency and selectivity of the various stationary phases are best evaluated by a test mixture separated under a standard set of conditions.

B.   DESCRIPTION OF STATIONARY PHASES

1.   Liquid-Liquid Partition

β,β'-Oxydipropionitrile.   β,β'-Oxydipropionitrile (BOP) is one of the most useful of the liquid stationary phases.

Successful applications for BOP include separations of aromatic amines, phenols, peroxides, aromatic alcohols, carbamates, and steroids. Retention times for sample components which are too strongly retained when a hydrocarbon mobile phase is used can be shortened by the addition of up to 10% chloroform or tetrahydrofuran.

Trimethylene Glycol. This stationary phase is also used with hydrocarbon carriers which can be satisfactorily modified with polar solvents. The most useful modifiers are tetrahydrofuran, dioxane, and chloroform. Alcohol modifiers should be avoided whenever possible, as they tend to strip the stationary phase. Trimethylene glycol is sufficiently selective for alcohols so that the separation of methanol, ethanol, propanol, isopropanol, butanol, isobutanol, and tertiary butanol is possible.

In comparison to β,β'-oxydipropionitrile, trimethylene glycol generally is more selective for alcohols and hydroxy-substituted compounds, peroxides, hydroperoxides, and ethylene oxide adducts, and less selective for amines and sulfamide compounds. A useful test for determining whether any degradation of performance has occurred over a period of time with an individual column involves the separation of two aromatic alcohols. The mixture employed consists, in order of elution, of α,α-dimethyl benzyl alcohol, α-methyl benzyl alcohol, 2-phenyl ethyl alcohol, cinnamyl alcohol, and benzyl alcohol, and was selected because it has compounds with both small and moderate k' values. In addition, it contains a closely eluting pair, cinnamyl alcohol and benzyl alcohol. Figure 11.1 compares the selectivity of these columns on the basis of this standard test mixture.

Carbowax®.* These materials are polyethylene glycols with molecular weights which vary from 200 to 20,000. The selectivity of the various Carbowax® stationary phases depends on the molecular weight of the polymer. High-molecular-weight Carbowax® 20M behaves in liquid chromatography as a weakly polar stationary phase. In gas chromatography, however, Carbowax® 20M is usually

---

*Registered trademark of Union Carbide.

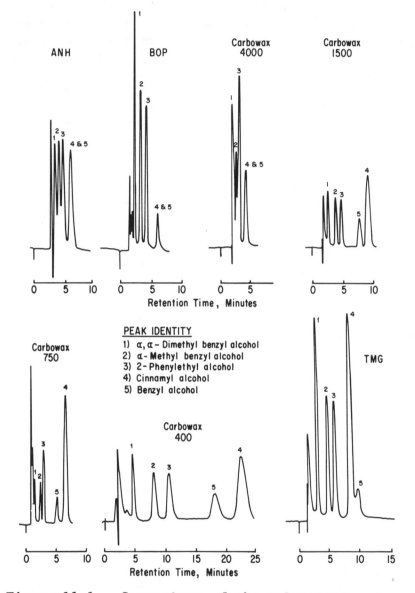

Figure 11.1. Comparison of the selectivity of seven stationary phases. <u>Conditions</u> - columns: Zipax® coated with 1% stationary phase, 1 m x 2.1 mm; mobile phase: <u>n</u>-hexane; flow rate: 1 ml/min.

considered a moderately polar stationary phase. The
Carbowaxes® also vary in physical form, from solids at
higher molecular weights to liquids at the lowest molecular weights. As a result, the efficiency obtained with
the various Carbowax® stationary phases will depend on
the molecular weight; the lower the molecular weight of
the polymer, the higher the efficiency of the column. In
addition to the efficiency differences observed, selectivity will also change as a function of molecular weight.

In general, Carbowax® 4000 is the highest-molecular-
weight polyethylene glycol that has been satisfactorily
used in liquid chromatography with hydrocarbon mobile
phases. The mobile phase can be modified with up to about
10% of polar modifiers such as tetrahydrofuran and dioxane. As described in Chapter 5, the mobile phase must be
presaturated with the stationary phase, and a precolumn
is required.

Carbowax® stationary phases are commonly used for the
separation of alcohols, steroids, and amines. The resolution between closely eluting compounds can be increased
by the use of a lower-molecular-weight Carbowax® and decreased by employing a higher-molecular-weight Carbowax®.
This ability to change resolution by altering the molecular weight used makes these polymers an interesting group
of stationary phases; however, some difficulties have
been encountered, such as short column life and problems
in homogeneously coating the solid polymers on the support.

Cyanoethyl Silicone ("ANH"). This stationary phase is
a silicone substituted with 50% cyanoethyl groups. It
is unique in that it will perform in both conventional
liquid-liquid partition and reversed-phase chromatography.
The reversed-phase applications will be discussed in Section B.2.

With hydrocarbon mobile phases, columns with cyanoethyl silicone as stationary phase are less selective for
the alcohol test mixture than columns with the phases
previously discussed. Polar solvents such as tetrahydrofuran, isopropanol, and chloroform are used as modifiers.
This stationary phase is soluble in chloroform.

Therefore, if chloroform is used as a modifier, the column must be operated at ambient temperature with a maximum of 1-2% modifier, and extreme care is required. The cyanoethyl silicone column is useful for halogen-substituted steroids, sulfamyl compounds, and substituted amines.

In Figure 11.1, the selectivities of the liquid-liquid partition stationary phases described are compared using the aromatic alcohol test mixture. In this study, a 1 meter x 2.1 mm column packed with Zipax® chromatographic support coated with 1% stationary phase was used. The flow of the mobile phase, n-heptane, was adjusted to 1 ml/min. The instrument used was a Du Pont model 820 equipped with a precision UV photometer detector.

From the comparative separations in Figure 11.1, it is apparent that the lowest-molecular-weight stationary phase produces the greatest resolution of the test mixture. Cyanoethyl silicone and Carbowax® 4000 stationary phases are essentially equivalent in resolving the test mixture. There is only a small difference between Carbowax® 1500 and Carbowax® 750. Trimethylene glycol shows an interesting selectivity in the reversal of the elution order of benzyl alcohol and cinnamyl alcohol. Knowing these differences in resolution and selectivity for the alcohol test mixture, one can facilitate the separation of closely related compounds by means of hydroxy substitutions to fit the particular analytical requirements.

To explore the selectivities and resolution capabilities of these stationary phases further, a polyether alcohol surfactant, Triton®* X-100, was chromatographed on each column. With Carbowax® 4000 and cyanoethyl silicone columns, the mixture was unretained and eluted with the solvent front. Both Carbowax® 1500 and Carbowax® 750 produced partial resolution of the constituents, while Carbowax® 400 gave the chromatogram shown in Figure 11.2. With this mixture, β,β'-oxydipropionitrile did not separate the first five peaks and gave only partial

---

*Registered trademark of Rohm & Haas.

separation of the remaining compounds. Trimethylene glycol produced a number of differences in the early portion
of the chromatogram, but with generally poor resolution.

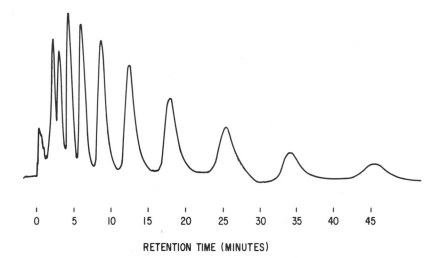

RETENTION TIME (MINUTES)

Figure 11.2. Separation of the constituents of
Triton X-100 (a polyphenyl ether detergent).
Conditions - column: 1% Carbowax® 400 on Zipax®,
1 m x 2.1 mm i.d.; mobile phase: n-hexane; column
pressure: 400 psi; flow rate: 1 ml/min; detector:
UV, 254 nm.

This comparison of stationary phases is not meant to
suggest that those discussed are the only ones useful in
liquid-liquid partition chromatography. Rather, the in-
tention is to illustrate the selectivities that have been
obtained.

2. Reversed-Phase Partition

Reversed-phase partitioning systems consist of a nonpolar stationary phase and a polar mobile phase. The
stationary phase may be a nonpolar liquid, a solid (such
as a hydrocarbon polymer), or a permanently bonded stationary phase. Mobile phases are usually combinations of
water and alcohols or water and acetonitrile. Resolution
and retention are governed by the percentage of alcohol

in the water-alcohol systems. Larger concentrations of
alcohol or acetonitrile tend to decrease retention times,
and larger percentages of water increase retention times.

Reversed-phase systems are generally less efficient
than other liquid-liquid systems because of the slower
diffusion of solutes in the more polar (and more viscous)
mobile phases. In addition, the difficulty in getting a
uniform coating of the polymeric stationary phases on the
solid support also limits column efficiency. To improve
the efficiency of the reversed-phase system, the mobile
phase producing the lowest viscosity should be selected;
that is, the use of methanol as a modifier is preferable
to the use of isopropanol. Column temperature should also
be increased above ambient when possible to improve mass
transfer. Unfortunately, operation at above ambient temperatures will reduce column life when liquids or non-
bonded polymer stationary phases are used.

Reversed-phase systems find their greatest applications for compounds that have hydrocarbon character and
are sparingly soluble in water. As a general rule, if a
compound is unretained on β,β'-oxydipropionitrile when a
hydrocarbon mobile phase is used, one should explore a
reversed-phase system. If a mixture can be chromatographed on β,β'-oxydipropionitrile but requires 5-10% of
a polar modifier in the mobile phase, it will probably
also chromatograph successfully on a reversed-phase system, but with a change in the elution order. The estrogens can be cited as an example. These compounds may be
satisfactorily separated on a β,β'-oxydipropionitrile
column with 5-10% tetrahydrofuran-heptane mobile phase.
Depending on the particular nonpolar stationary phase,
the estrogens also separate on a reversed-phase system
in 20-30% methanol-water. The ability to analyze a mixture with two systems is particularly useful when a complex sample matrix is involved.

Hydrocarbon Polymer. This stationary phase is a nonpolar saturated hydrocarbon polymer. It was first applied
to the separation of fused-ring aromatics and is generally useful for the analysis of compounds with considerable
aromatic character. Because of both the difficulty of

coating the stationary phase on the solid support and a
limited temperature range, this system produces separations of somewhat lower efficiency. However, the hydrocarbon polymer stationary phase, even with its low efficiency, has sufficient selectivity to separate mixtures
that cannot be analyzed on other liquid-liquid systems.
Water-methanol mobile phases are preferred, with retention times adjusted by changing the percentage of methanol.

Cyanoethyl Silicone. Cyanoethyl silicone was described previously as a polar stationary phase; however,
many of the useful separations achieved with cyanoethyl
silicone are reversed-phase applications in which this
material serves as a relatively nonpolar stationary phase.
Cyanoethyl silicone usually produces columns of higher
efficiency than those obtained with hydrocarbon polymer,
probably because the silicone coats more evenly on the
solid support. This stationary phase is limited in the
useful temperature range, ambient to 40°C, and in the
percentage of alcohol modifier (up to 50% methanol). Although many of the separations achieved on cyanoethyl
silicone can be accomplished also on hydrocarbon polymer,
the former offers advantages in some areas, particularly
with phenolic and halogen- or amine-substituted compounds.
This stationary phase is also useful for the separation
of straight-chain hydrocarbons, using water-acetonitrile
mixtures as the mobile phase.

Octadecyl Permaphase®*. Octadecyl (ODS) Permaphase®
is a permanently bonded silicone based on the chemistry
described by Kirkland (5). This stationary phase has
the advantage of uniform coating on the solid support and
thermal stability in a wide variety of mobile phases.
Because of its hydrocarbon character, its main utility is
in a reversed-phase application. With the ability to increase column temperature and thereby reduce the viscosity of the mobile phase, one may achieve column efficiencies 4-5 times higher than those of some of the other
reversed-phase systems. For this reason, ODS Permaphase®

---

*Du Pont trademark.

is preferred over hydrocarbon polymer and cyanoethyl silicone for many reversed-phase applications.

A comparison of separations of substituted anthraquinones on columns packed with hydrocarbon polymer and with
ODS Permaphase® is shown in Figure 11.3. These chromatograms illustrate both the increased efficiency and the
greater retentivity of ODS Permaphase®. Although part of
the increased efficiency is due in this case to the elevated temperatures employed, a comparison of the two stationary phases at the same temperature also shows that
ODS Permaphase® has superior performance over the non-
bonded polymers. This improved performance is probably
due to a more uniform layer of stationary phase on the
solid support. The mobile phase was identical; however,
there was a substantial difference in flow (0.8 ml/min
versus 2.2 ml/min) and in temperature (ambient versus
55°C). The difference in relative peak heights is not
significant, as two test solutions having different concentrations of constituents were used in the evaluation
of the columns.

3.   Ion-Exchange

The speed and efficiency of ion-exchange separations
have been greatly increased by the development of packings
consisting of a thin layer of ion-exchange resin coated
on the surface of a solid-core support. This approach
has been described by Kirkland (6), as well as by Horvath
et al. (7). These packings have a smaller capacity than
the conventional ion-exchange resins but offer the advantages of rapid equilibration with the mobile phase and
high-speed separations. They can also be dry-packed in a
column, as swelling of the resin is negligible. The limited capacity of the resin requires small sample sizes,
however, and sensitive detectors to avoid overloading
the column.

The separations described here were achieved using an
ion-exchange resin coated on the surface of Zipax®. General applications of strong anion-exchange (methacrylate
polymer substituted with quaternary ammonium groups) and
strong cation-exchange (sulfonated fluorocarbon polymer)

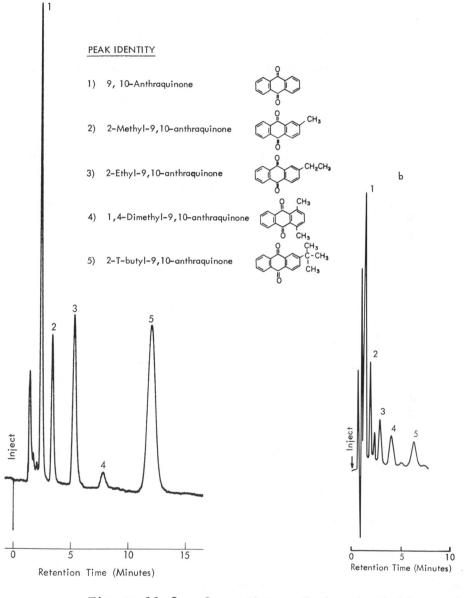

PEAK IDENTITY

1) 9, 10-Anthraquinone

2) 2-Methyl-9,10-anthraquinone

3) 2-Ethyl-9,10-anthraquinone

4) 1,4-Dimethyl-9,10-anthraquinone

5) 2-T-butyl-9,10-anthraquinone

Retention Time (Minutes)

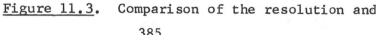

Figure 11.3. Comparison of the resolution and

Figure 11.3 (cont.). efficiency of hydrocarbon polymer and ODS Permaphase® stationary phases. Conditions - (a) ODS Permaphase®. Mobile phase: 40%
water-60% methanol (v/v); column pressure: 1200
psig; temperature: 40°C; flow rate: 2.2 ml/min.
(b) Hydrocarbon polymer. Mobile phase: 50% water-
50% methanol (v/v); column pressure: 900 psig;
temperature: ambient; flow rate: 0.8 ml/min.

resins coated on Zipax® chromatographic support have been
described previously (8,9,10). In addition to the ion-
exchange mechanism, other forms of retention may occur
with these polymeric phases.

Aqueous mobile phases containing various ionic components are used for ion-exchange chromatography, with
resolution and retention varied by changes in pH and ionic
strength. Small amounts of alcohol may also be added to
the mobile phase to reduce the retention times of strongly retained compounds. The efficiency of the superficially porous or pellicular ion-exchange column (H = 1-2 mm
at mobile-phase velocities of 1-2 cm/sec) is less than
that realized with other types of columns; some improvement in efficiency can be realized, however, by increasing the column temperature and thus improving mass
transfer.

C. APPLICATIONS OF LIQUID CHROMATOGRAPHY

In this section, specific separations carried out in
the Du Pont Instruments Applications Laboratory are discussed. Insofar as possible, the samples illustrated
are typical of a particular class of compounds or field
of research.

1. Analgesics

High-speed liquid chromatography has been applied to
the analysis of both single- and multiple-component analgesic tablets (11,12), as well as to the analysis of free
salicylic acid in aspirin-based analgesics. The active
ingredients are chromatographed on a strong anion-
exchange column with a mobile phase consisting of distilled water adjusted to a pH of 9.2 with a borate buffer,

and the ionic strength adjusted by the addition of 0.002<u>M</u> sodium nitrate. Increasing the ionic strength of the carrier with sodium nitrate reduces the time required for the analysis without substantially impairing the resolution of the components.

A typical separation of a multiple-component analgesic is shown in Figure 11.4. Benzoic acid elutes in a

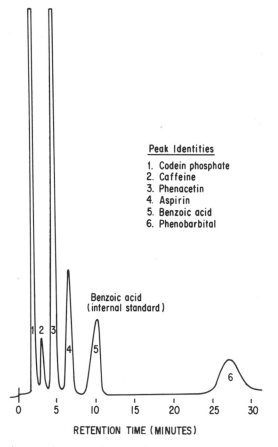

Peak Identities

1. Codein phosphate
2. Caffeine
3. Phenacetin
4. Aspirin
5. Benzoic acid
6. Phenobarbital

Benzoic acid
(internal standard)

RETENTION TIME (MINUTES)

<u>Figure 11.4</u>. Separation of a multicomponent analgesic tablet. <u>Conditions</u> - column: strong anion exchange (SAX); mobile phase: aqueous pH 9.2 + 0.005 <u>M</u> NaNO$_3$; column pressure: 1200 psig; flow rate: 1.2 ml/min; detector: UV photometer at 254 nm.

convenient region of the chromatogram and can be used as
an internal standard for quantitative analyses. This
compound is well resolved from other tablet constituents
and does not increase the time required for the analysis.
By employing the internal standard method and an electronic integrator, a precision of $\pm 1\%$ relative can be
achieved.

A strong cation-exchange column can also be used for
the analysis of analgesics, although a change in the elution order of the constituents can be expected. For
example, an anion-exchange column elutes a typical analgesic in the following order: caffeine, phenacetin,
aspirin, whereas a cation-exchange column elutes these
compounds in the order aspirin, caffeine, phenacetin.
Although this change in elution order can be important
in trace analysis, the cation-exchange resin is not selective enough to have wide applicability for analgesics.

Under the conditions given for the separation in Figure 11.4, the free salicylic acid is strongly retained
by the column. By increasing the ionic strength of the
mobile phase to at least 0.05$\underline{M}$ sodium nitrate, free
salicylic acid can be eluted. Under these conditions,
the active ingredients are unretained on the column and
elute with the solvent front.

A typical chromatogram of free salicylic acid in a
multiple-component analgesic tablet is shown in Figure
11.5.

2. Aromatic Carboxylic Acids

Aromatic carboxylic acids have been analyzed in a number of matrices, using the strong anion-exchange column
previously described. An example is shown in Figure
11.6. The choice of the mobile-phase pH depends on the
particular problem. A change of pH will alter the elution order of some of these acids, and this change in
elution order is important both for compound identification and in trace analysis. In trace analysis it is
desirable to have the minor component elute before the
major component. Usually the sample must be injected in
relatively high concentration to detect the minor

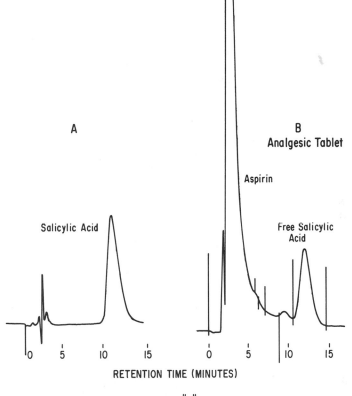

A

B
Analgesic Tablet

Aspirin

Salicylic Acid

Free Salicylic
Acid

RETENTION TIME (MINUTES)

NOTE: Spikes on chromatogram "B" are electronic
integrator event marks

Figure 11.5. Analysis of free salicylic acid in
a multicomponent analgesic tablet.

constituent. In many cases, the column is overloaded by
the major compound, resulting in a tailing peak with accompanying difficulties in measuring small peaks on the
trailing edge.

In addition to the pH, the concentration and the type
of modifying ion are important in affecting a desired
separation. The relative effect of a particular ionic
species can be studied at a given pH by adding increasing

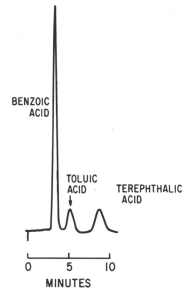

Figure 11.6.   Separation of a mixture of aromatic
carboxylic acids.   Conditions - column: strong anion exchange; mobile phase: aqueous pH 9.18 buffer
+ 0.02 $\underline{M}$ NaNO3; column pressure: 1200 psig; flow
rate: 1.2 ml/min; detector: UV photometer.

amounts of the salt under investigation until a predetermined k' value for a test compound (or mixture) is obtained.   When the test mixture shown in Figure 11.6 is
used with an aqueous carrier at a pH of 9.2 and a concentration of NaClO4 required to produce unity k' values,
the relative increases in concentration for some common
salts compared to the perchlorate are as follows: NaNO3,
5; Na2HPO4, 80; and Na2SO4, 100.   Because of the large
concentrations of the phosphate and sulfate required,
these salts are less useful as ionic strength modifiers.
Phosphates are also useful in the pH adjustment of the
mobile phase.   The halide salts corrode stainless steel
and therefore should be avoided as ionic strength modifiers.   There has been no evidence in this laboratory of
the corrosion of stainless steel when perchlorate salts

are used.

3. Hydrocarbons

Fused-Ring Aromatics. The use of reversed-phase partition chromatography for the separation of fused-ring
aromatics was described in Chapter 5. The recent development of ODS Permaphase® has substantially improved the
efficiency and resolution obtainable for these compounds
in a number of matrices. Figure 11.7a shows the separation of a synthetic mixture of fused-ring aromatics. A
chromatogram of a benzene extract of an air filter sample
is shown in Figure 11.7b.

The resolution and retention of the aromatics are governed by the percentage of alcohol in the mobile phase.
When the alcohol is increased to 90%, all of the compounds
elute as a single peak. This technique provides a rapid
estimation of the total aromatics. Should a more selective analysis be required, a reduction of alcohol content
in the mobile phase allows the measurement of the individual compounds, as shown in Figure 11.7.

Straight-Chain Hydrocarbons. The application of liquid
chromatography for the analysis of hydrocarbons was reported by Locke (13), who used squalane as a stationary
phase and acetonitrile as a mobile phase. The range of
hydrocarbons that can be analyzed with this system can
be extended by modifying the mobile phase with water. A
typical hydrocarbon separation is shown in Figure 11.8.

Substituted Aromatics. The system consisting of ODS
Permaphase® and a water-alcohol mobile phase will also
resolve halogen-substituted benzenes, phenyl homologs,
and substituted biphenyls under conditions similar to
those employed for the separation of fused-ring aromatics.
A separation of benzene and nine chlorinated derivatives
is shown in Figure 11.9. In addition to these compounds,
bromobenzenes, iodobenzenes, and mixed halogen derivatives have been chromatographed with this system.

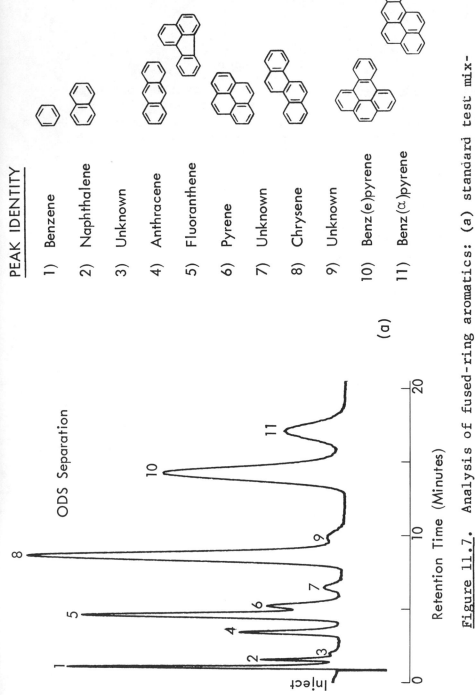

PEAK IDENTITY

1) Benzene

2) Naphthalene

3) Unknown

4) Anthracene

5) Fluoranthene

6) Pyrene

7) Unknown

8) Chrysene

9) Unknown

10) Benz(e)pyrene

11) Benz(α)pyrene

ODS Separation

(a)

Retention Time (Minutes)

Figure 11.7. Analysis of fused-ring aromatics: (a) standard test mixture; (b) benzene extract of an air-filter sample. Conditions - column: ODS Permaphase®, 1 m x 2.1 mm i.d.; mobile phase: 60% methanol-40% water; column temperature: 50°C; column pressure: 1200 psig; flow rate: 2.1

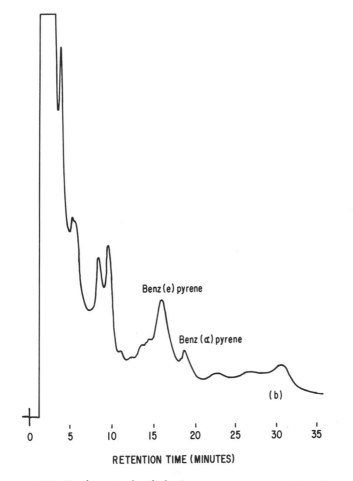

Benz (e) pyrene

Benz (α) pyrene

(b)

RETENTION TIME (MINUTES)

<u>Figure 11.7</u> (cont.) (b) benzene extract of an air-filter sample.

4. <u>Steroids</u>

Steroids have been satisfactorily analyzed on both types of liquid-liquid partition systems. In many cases either system can be employed to separate the same mixture of compounds. The ability to change the chromatographic system is particularly important in the analysis

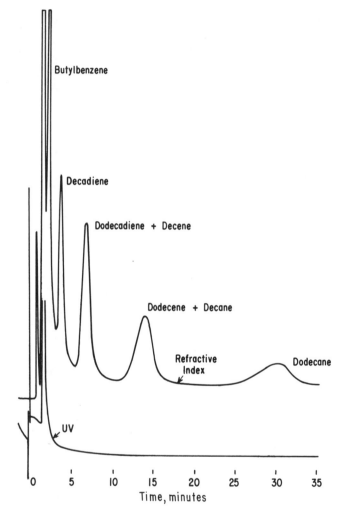

<u>Figure 11.8.</u> Separation of aromatics, paraffins, olefins, and diolefins. <u>Conditions</u> - column: 1% squalane on Zipax®, 1 m x 2.1 mm i.d.; mobile phase: acetonitrile-20% water; column pressure: 1200 psig; flow rate - 1.2 ml/min; detector: UV at 254 nm and refractive index.

PEAK IDENTITY

1) Benzene
2) Monochlorobenzene
3) Ortho-dichlorobenzene
4) 1,2,3-trichlorobenzene
5) 1,3,5-trichlorobenzene
6) 1,2,4-trichlorobenzene
7) 1,2,3,4-tetrachlorobenzene
8) 1,2,4,5-tetrachlorobenzene
9) Pentachlorobenzene
10) Hexachlorobenzene

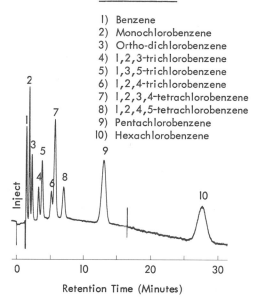

Retention Time (Minutes)

Figure 11.9. Separation of a mixture of chlorinated benzenes. Conditions - column: ODS Permaphase®,
1 m x 2.1 mm i.d.; mobile phase: 50% water-50%
methanol (v/v); temperature: 60°C; pressure: 1200
psig; flow rate: 2 ml/min.

of steroids in oils, ointments, and creams, in which the
formulation constituents may interfere with the analysis
of steroids in low concentrations. The most successful
liquid-liquid system for steroids used in this laboratory
has consisted of columns with β,β'-oxydipropionitrile as
the stationary phase and n-heptane and tetrahydrofuran-
modified n-heptane as a mobile phase. This system has
been particularly valuable for the analysis of estrogens
and substituted estrogens. The analysis of estradiol in
a sesame oil-based injectable is shown in Figure 11.10.
    Prednisolone and prednisolone derivatives can be ana-
lyzed by liquid-liquid systems. The particular system
selected will depend on the specific mixture and the requirements of the analysis. If the components differ

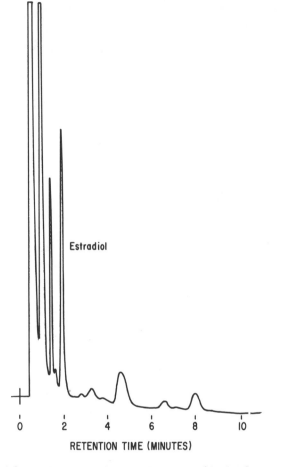

Figure 11.10. Separation of estradiol from sesame oil in an oil-based injectable. <u>Conditions</u> - column: 1% β,β'-oxydipropionitrile on Zipax®, 1 m x 2.1 mm i.d.; mobile phase: 5% tetrahydrofuran in n-heptane; column pressure: 600 psig; flow rate: 1 ml/min; detector: UV at 254 nm.

primarily in the position or the number of hydroxyl substitutions, a liquid-liquid system involving a polar stationary phase and a nonpolar mobile phase is usually the

best choice. If the major differences are in the side
chain, a reversed-phase system is preferred.

Examples of both types of separations are illustrated
in Figures 11.11 and 11.12. In addition to the two general types shown, other, more polar steroids have also
been chromatographed. These include steroid salts, common constituents of pharmaceutical formulations, and

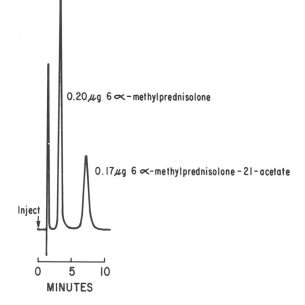

Figure 11.11. Separation of 6 α-prednisolone-21-
acetate from 6 α-methyl prednisolone-21-acetate.
Conditions - column: hydrocarbon polymer on Zipax®,
1 m x 2.1 mm i.d.; mobile phase: 15% methanol in
water; column pressure: 1200 psig; flow rate: 1.2
ml/min; detector: UV photometer.

polyhydroxy steroids, shown in Figure 11.13. Improvements in the resolution of these compounds could be expected if they were chromatographed with columns using
trimethylene glycol or Carbowax® 400 as stationary
phases. Both of these liquids have greater selectivity
for the hydroxy function.

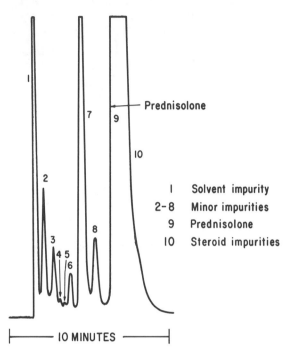

<figure>
Prednisolone

1   Solvent impurity
2-8 Minor impurities
9   Prednisolone
10   Steroid impurities

├──── 10 MINUTES ────┤
</figure>

<u>Figure 11.12.</u> Separation of prednisolone from im-
purities. <u>Conditions</u> - column: Zipax® coated with
1% ethylene glycol; mobile phase: <u>n</u>-heptane + 2%
chloroform; column pressure: 400 psig; flow rate:
1 ml/min.

Liquid chromatography has been successfully applied
also to the analysis of halogenated steroids, particularly the fluoro derivatives, where the selectivity is
sufficient to distinguish between α- and β-substitutions.
Further advances in steroid analysis can be expected,
especially in regard to the conjugates.

5. Herbicides

Herbicides of the substituted urea, chlorophenoxy ace-
tic acid, phenoxy acid, and triazine types have been successfully separated. The choice of the chromatographic
system will depend on the solubility of the compound and

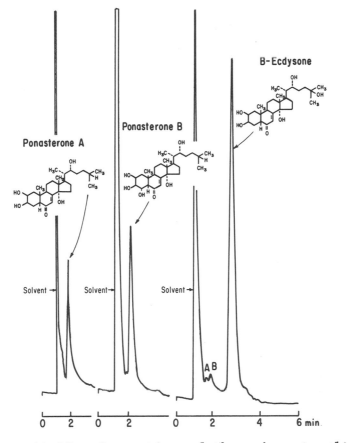

<u>Figure 11.13.</u>   Separation of three insect-molting hormones. <u>Conditions</u> - column: 1% β,β'-oxydipro- pionitrile on Zipax®, 1 m x 2.1 mm i.d.; mobile phase: 10% tetrahydrofuran in <u>n</u>-heptane; column pressure: 600 psig; flow rate: 1 ml/min.

its functionality. Examples involving two types of her- bicides and two chromatographic systems are discussed below. Although herbicides vary greatly in chemical type, the examples should serve as a starting point for the selection of a chromatographic system suitable for other specific compounds.

Substituted Urea Herbicides. The substituted urea herbicides have been chromatographed on a liquid-liquid partition system consisting of a β,β'-oxydipropionitrile
stationary phase and n-heptane or n-heptane with 5-10% of
a polar modifier such as tetrahydrofuran or chloroform as
a mobile phase. An example of this type of separation is
shown in Figure 11.14. In this example, samples of the

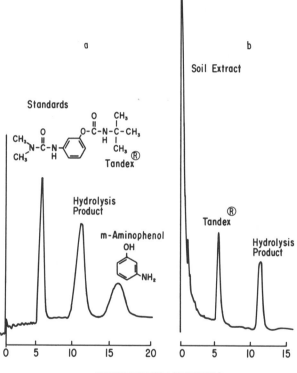

RETENTION TIME (MINUTES)

Figure 11.14. Tandex® and related compounds. (a)
Tandex® hydrolysis product, and reaction intermediate; (b) Tandex® and hydrolysis product from a soil
extract. Conditions - Same as for Figure 11.13,
except 5% tetrahydrofuran in n-heptane; flow rate:
1.4 cc/min; column pressure: 840 psig.

parent compound, Tandex®*, the intermediate, m-amino-
phenol, and the hydrolysis product, a urea phenol, were
resolved as standards. Both the parent compound and the
hydrolysis product were detected in a soil extract. Although there was some background in the soil extract, it
did not interfere with the analysis of the herbicide and
its hydrolysis product. Other urea-substituted herbicide
separations employing a β,β'-oxydipropionitrile stationary phase have been reported (7).

Phenoxyacetic Acids. Esters of phenoxyacetic acid-
based herbicides can be separated on ODS Permaphase® with
a 40% water-60% methanol mobile phase. The separation of
a synthetic mixture of 2,4-D esters is shown in Figure
11.15. Under these conditions, the free acids are unretained and elute with the solvent front. The free acids
can be determined in the reversed-phase system by reducing the percentage of methanol and adding dilute phosphoric acid, which reduces the ionization of the acids
and produces more symmetrical peaks. The free acids can
also be chromatographed on a column with ODS Permaphase®
as the packing and n-heptane-5% isopropanol as the mobile
phase. A strong anion-exchange column should also prove
useful in the separation of the free acids.

The phenoxy acid-based herbicides may be separated
with a system similar to that employed for 2,4-D mixtures.
The 2,4,5-T esters have longer retention times than the
corresponding 2,4-D esters. The retention time of either
type lengthens as the chain length of the ester increases.
A strong anion-exchange column has also been used for the
analysis of these compounds.

Triazine herbicides have been chromatographed on
liquid-liquid systems with β,β'-oxydipropionitrile or
trimethylene glycol as a stationary phase and n-heptane
or modified n-heptane as the mobile phase.

6. Pesticides

Liquid chromatography should prove to be a valuable
adjunct to gas chromatography as a confirmatory tool for

*Registered trademark of Niagara Chemical Company.

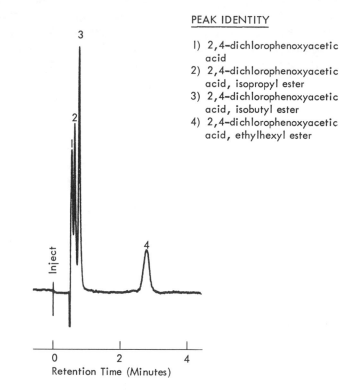

PEAK IDENTITY

1) 2,4-dichlorophenoxyacetic acid
2) 2,4-dichlorophenoxyacetic acid, isopropyl ester
3) 2,4-dichlorophenoxyacetic acid, isobutyl ester
4) 2,4-dichlorophenoxyacetic acid, ethylhexyl ester

Figure 11.15. Separation of 2,4-D and 2,4-D esters. **Conditions** - column: ODS Permaphase®; mobile phase: 60% methanol-40% water (v/v); pressure: 1200 psig; temperature: 60°C; flow rate: 2.2 ml/min.

pesticides which can be analyzed by gas chromatography, for the rapid analysis of pesticides which require the synthesis of derivatives for analysis, and for the analysis of pesticides that cannot be handled by gas chromatography. Both types of liquid-liquid systems have been employed for pesticide analysis. The two most useful liquid stationary phases have been β,β'-oxydipropionitrile and trimethylene glycol, with n-heptane or n-heptane and 5-10% of a polar modifier as the mobile phase. Most commonly, tetrahydrofuran, dioxane, and chloroform have been selected as the modifier.

Organophosphorus Pesticides. An example of the analysis of an organophosphorus pesticide is shown in Figure
11.16. Both the technical-grade larvicide Abate®* and a
chloroform extract of a salt-water marsh pond sample are
illustrated (14). To obtain the latter chromatogram, a

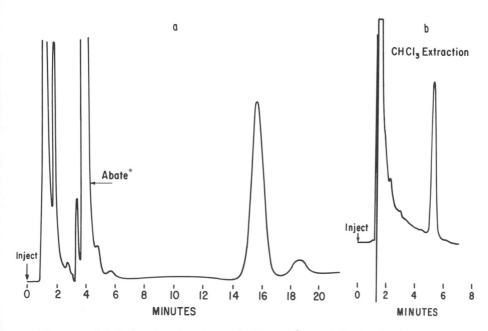

Figure 11.16. Analysis of Abate®. (a) Technical-
grade Abate®; (b) detection of Abate® in a chloroform extract of a sample from a salt-water pond.
Conditions - column: 1% β,β'-oxydipropionitrile on
Zipax®, 1 m x 2.1 mm i.d.; mobile phase: n-heptane;
column pressure: 600 psig; flow rate: 1 ml/min;
detector: UV photometer; sample: 5 μl 0.5% CHCl₃
solution.

salt-water pond sample was extracted with chloroform and
concentrated 100-fold by evaporation. The final concentration of the sample in Figure 11.16 is 2 ppm. The

*Registered trademark of American Cyanamid Company.

minimum detectable level of this larvicide is approximately 0.1 ppm. Other related organophosphorus compounds have also been chromatographed by employing similar systems.

Carbamates. The carbamates have also been analyzed directly by liquid chromatography without derivative preparation. In many cases, these compounds can be chromatographed on both types of liquid-liquid systems. An example of the analysis of a carbamate and its major decomposition products is shown in Figure 11.17, which

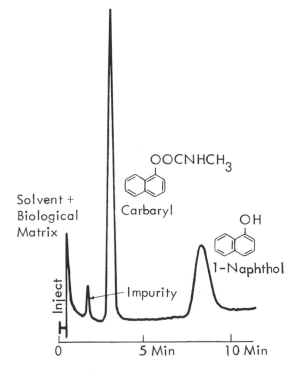

Figure 11.17. Analysis of carbaryl and its major decomposition product in a plant extract. Conditions - column: 1 m x 2.1 mm i.d.; 1% trimethylene glycol on Zipax®; mobile phase: n-hexane; column pressure: 1200 psig; flow rate: 2 ml/min; detector: UV at 254 nm.

affords a comparison of the selectivity of β,β'-oxydipro-
pionitrile and trimethylene glycol columns. Although
both stationary phases will accomplish the separation,
the trimethylene glycol column is more selective for the
hydroxyl group and produces much greater resolution between carbaryl and 1-naphthol. The background from the
plant extract elutes with the solvent front in both cases
and does not interfere with the determination of carbaryl
or 1-naphthol. The same separation has also been accomplished by a reversed-phase method, using a 10% methanol-
90% water mobile phase and ODS Permaphase®. Because of
the vast differences in solubility of the various carbamates, both systems should find use for this analysis.
However, the ability to vary the mobile phase without
eluting the stationary phase makes the ODS Permaphase®
column particularly useful in carbamate analyses.

Miscellaneous Pesticides. In addition to these pesticide types, many of the chlorinated hydrocarbons have
also been chromatographed. These separations have been
carried out with β,β'-oxydipropionitrile/heptane-chloro-
form or ODS Permaphase®/methanol-water systems. The main
application of liquid chromatography for the analysis of
chlorinated pesticides is as a confirmatory tool for gas
chromatography, or as an alternative technique when the
matrix encountered requires elaborate cleanup procedures
before gas chromatographic analysis is feasible.

The analysis of methoxychlor can serve as an example
of the potential use of both liquid chromatographic and
gas chromatographic techniques. Methoxychlor is a late-
eluting peak on most gas chromatographic column systems.
Sample matrix constituents tend to elute in the same
region of the gas chromatogram as methoxychlor, causing
some confusion as to the identity of the compounds.
These interferences are particularly common with residue
samples from fatty tissues. With a liquid-liquid partition system consisting of β,β'-oxydipropionitrile as a
stationary phase and n-heptane—5% tetrahydrofuran,
methoxychlor elutes early in the chromatogram. Under
these conditions, the sample background usually does not
interfere, as it either elutes before the methoxychlor
or is not a UV absorber.

7. Hydroquinones and Related Compounds

The hydroquinones are generally freely soluble in
water or water-methanol solvents, ruling out the use of
reversed-phase methods for the analysis of these compounds. A number of stationary phases may be selected
when the molecules are dihydroxy substituted. However,
the strong interactions of the hydroxyl groups with Car-
bowax® below molecular weight 4000, and with trimethylene
glycol, generally preclude the use of these columns. The
best separations are carried out with a β,β'-oxydipropio-
nitrile/n-heptane—10% tetrahydrofuran system. As the
side-chain substitutions increase in chain length and the
compounds become more hydrocarbon soluble, the retention
times decrease so that the 2,5-di-t-butyl derivative is
unretained in n-heptane mobile phase. In this case,
either reversed-phase or trimethylene glycol/n-heptane
may be used satisfactorily as a chromatographic system.

In Figure 11.18, the separation of a mixture of hydroquinone and related compounds is illustrated. This system has been used to separate both the t-butyl and the
di-t-butyl derivatives, as well as the o-, m-, and p-
isomers.

8. Antibiotics

In addition to the rifampin samples illustrated in
Chapter 5, liquid chromatography has also been applied to
the analysis of penicillin. The penicillins vary widely
in solubility and in the substitutions on the molecule.
It is doubtful that any single chromatographic system
will be applicable to all of the penicillin derivatives.
At this writing, high-speed liquid chromatography has been
most successful for the analysis of the methyl benzyl
esters of the penicillins. Further advances in the determination of the penicillins can be expected as more work
is devoted to the applications of these compounds. A
polyamide stationary phase with an ethanol-5% hexane mobile phase was developed for penicillin analysis. The
elution time of the methyl benzyl esters is governed by
the percentage of hexane in the mobile phase. The separation of the methyl benzyl ester of penicillin G from its

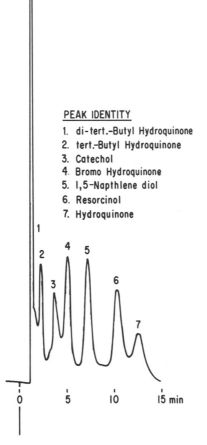

PEAK IDENTITY
1. di-tert.-Butyl Hydroquinone
2. tert.-Butyl Hydroquinone
3. Catechol
4. Bromo Hydroquinone
5. 1,5-Napthlene diol
6. Resorcinol
7. Hydroquinone

Figure 11.18. Separation of hydroquinone and related compounds. Conditions - same as for Figure
11.13.

impurities is shown in Figure 11.19.

The carbohydrate-containing antibiotics have also been
separated on a Carbowax® 750 column with a hexane-isopro-
panol mobile phase. The hydroxy groups on the molecule
interact with the Carbowax® to retain the sample. The

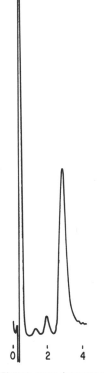

RETENTION TIME (MINUTES)

Figure 11.19. Separation of the methyl benzyl ester of penicillin from minor impurities. Conditions - column: polyamide on Zipax®, 1 m x 2.1 mm
i.d.; mobile phase: 5% n-hexane in ethanol; column
pressure: 900 psig; flow rate: 0.9 ml/min; detector: UV at 254 nm.

antibiotic is sufficiently soluble in isopropanol so that
small percentages added to the mobile phase will

satisfactorily elute the compounds. Retention times are governed by the percentage of isopropanol in the mobile phase.

9. Dyes and Dye Intermediates

Although some advances in the separations of dyes and dye intermediates have been made, a great deal remains to be accomplished. These compounds differ widely in solubility and in structure and, therefore, require a variety of liquid chromatographic systems for separation. This area of liquid chromatography applications can be expected to expand as new column technology becomes available and wider use is made of gradient elution techniques.

Naphthalene Sulfonic Acids. The naphthalene sulfonic acids are an important group of dye intermediates. These compounds have been chromatographed on a strong anion-exchange column with a mobile phase consisting of 0.01$\underline{M}$ aqueous $NaClO_4$. A typical separation is shown in Figure 11.20.

In addition to the intermediates, some of the finished dyes derived from naphthalene sulfonic acids have been chromatographed. In general, higher ionic strengths are required than for the intermediates, as well as temperatures above ambient.

Anthraquinones. The anthraquinone dyes and dye intermediates have been chromatographed with both types of liquid-liquid systems, the choice depending on the substitution on the molecule. If the substitution is an alkyl or aryl group, a reversed-phase system should be selected. With these substitutions, anthraquinone chromatographs satisfactorily using systems similar to those employed for the fused-ring aromatics; however, the ketone function increases the solubility of the molecule in alcohols. For this reason, smaller percentages of alcohol are used to modify the mobile phase, as compared to the corresponding fused-ring aromatic.

When the anthraquinone molecule is substituted with amino or hydroxyl groups, its behavior in the reversed-phase system changes markedly. This interaction and the subsequent retention by the column depend on both the

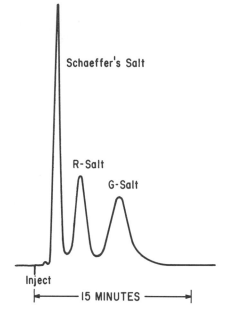

Figure 11.20. Separation of naphthalene sulfonic acid salts. <u>Conditions</u> - column: strong-anion exchange on Zipax®; mobile phase: distilled water + 0.025<u>M</u> NaClO₄; pressure: 1200 psig; flow rate: 1.2 ml/min.

number and the location of the substitution on the molecule. In the case of the hydroxy substitution, this interaction is particularly evident. The dihydroxyanthraquinones are a good example of the problems encountered. The 1,5- and 1,8-dihydroxyanthraquinones can be chromatographed with a system consisting of a stationary phase of ODS Permaphase® and a 15% methanol-85% water mobile phase. Under these conditions, 1,2-dihydroxyanthraquinone is completely retained. This compound can be eluted from the column with 1/1 methanol/water, but tails badly. The addition of dilute phosphoric acid to the mobile phase reduces the tailing; however, the peak is fairly broad. With the β,β'-oxydipropionitrile/heptane-tetrahydrofuran liquid-liquid system, all of the isomers produce symmetrical peaks but still exhibit very large differences in k' values.

An amino substitution on the anthraquinone molecule does not seem to produce the same problems. Aminoanthraquinone dyes have been successfully chromatographed on both types of liquid-liquid systems with little difficulty. An example of a typical analysis of an aminoanthraquinone dye on a liquid-liquid system is shown in Figure 11.21. When a reversed-phase system is employed for the analysis of the same dye, the elution order is reversed; the impurities are more strongly retained and elute after the parent compound.

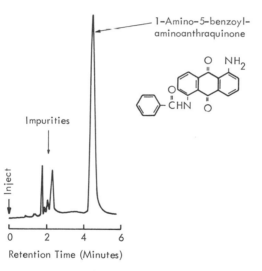

Figure 11.21. Separation of aminoanthraquinone dye from impurities. Conditions - Same as for Figure 11.13.

10. Coumarin and Derivatives

The coumarins constitute an example of naturally occurring phenolic compounds present in a large number of plant species. The coumarins are especially common in grasses, citrus fruits, legumes, and tobacco. These compounds may be chromatographed on a reversed-phase system consisting

of a cyanoethyl silicone stationary phase and a distilled
water mobile phase. Under these conditions, base-line
separation between coumarin and four coumarin derivatives
is achieved in less than 25 min, as shown in Figure 11.22.
Another liquid-liquid system, β,β'-oxydipropionitrile/n-
heptane—5% tetrahydrofuran, has also been employed for
this analysis with comparable resolution, but with a
changed elution order.

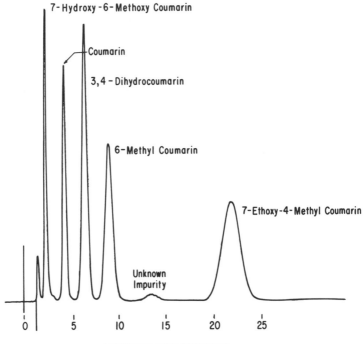

Figure 11.22. Separation of coumarin from several
derivatives. Conditions - column: 1% cyanoethyl
silicone on Zipax®, 1 m x 2.1 mm i.d.; mobile phase:
distilled water; column pressure: 1200 psig; flow
rate: 1.5 ml/min; column temperature: 40°C; detector: UV at 254 nm.

11.   Alkaloids

The alkaloids in the extracts of two plants, Veratrum
viride and ipecac, have been chromatographed. The ipecac
alkaloids, which are C28 and C29 five-ring molecules, are
very strongly retained on β,β'-oxydipropionitrile and require large additions of a polar modifier for elution.
On a cyanoethyl silicone column, these compounds are sufficiently retained to be separated from the solvent peak
and from each other. The separation of emetine and cepha-
eline, the two principal alkaloids found in ipecac, is
shown in Figure 11.23.

The Veratrum viride alkaloids are primarily C27 five-
and six-ring compounds with varying numbers of hydroxyl
substitutions. These alkaloids require a more selective
stationary phase than the ipecac alkaloids. Veratrum
viride extracts contain a greater number of individual
alkaloid compounds with small structural differences.
The β,β'-oxydipropionitrile column with n-heptane mobile
phase provides good resolution of the individual compounds without unduly long separation times. Should
greater resolution be required, either trimethylene glycol or Carbowax® 400 can be used as the stationary phase.

D.   CONCLUSIONS

The use of high-speed liquid chromatography can be
expected to grow rapidly in the future as further advances are made in column technology and more workers become active in the field. The applications illustrated
here provide a survey of some of the chemical classes
which have been analyzed to date in this laboratory. The
list should grow at a rapid rate with further developments of the chemically bonded stationary phases and the
wider use of gradient elution techniques.

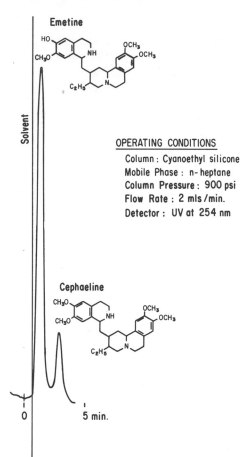

OPERATING CONDITIONS

Column : Cyanoethyl silicone
Mobile Phase : n-heptane
Column Pressure : 900 psi
Flow Rate : 2 mls/min.
Detector : UV at 254 nm

Figure 11.23. Separation of the principal alkaloids of an ipecac extract. Conditions - same as for Figure 11.22, except carrier: n-heptane; column pressure: 900 psig; flow rate: 2 ml/min.

References

1. R. E. Majors, Symposium on Advances in Chromatography, Miami Beach, Fla., June 2-5, 1970.
2. J. J. Kirkland, J. Chromatog. Sci., 7, 7 (1969).
3. J. J. Kirkland, U. S. Patent 3,505,785.
4. H. R. Felton, J. Chromatog. Sci., 7, 7 (1969).
5. J. J. Kirkland and J. J. DeStefano, J. Chromatog. Sci., 8, 309 (1970).

6.   J. J. Kirkland, J. Chromatog. Sci., $\underline{7}$, 361 (1969).
7.   G. G. Horvath, B. A. Preiss, and S. R. Lipsky, Anal.
     Chem., $\underline{39}$, 1422 (1967).
8.   J. J. Kirkland, J. Chromatog. Sci., $\underline{8}$, 72 (1970).
9.   J. A. Schmit and R. A. Henry, in press.
10. Chromatographic Methods, 820M1, 820M3, 820M5, Instru-
     ment Products Division, E. I. du Pont de Nemours &
     Co., Wilmington, Del.
11. R. A. Henry and J. A. Schmit, Chromatographia, $\underline{3}$,
     116 (1970).
12. J. A. Schmit, Pittsburgh Conference on Analytical
     Chemistry and Spectroscopy, Cleveland, Ohio, March
     1970.
13. D. C. Locke, J. Chromatog., $\underline{35}$, 24 (1968).
14. R. A. Henry, J. A. Schmit, J. F. Dieckman, and F. J.
     Murphy, 160th National Meeting of the American Chem-
     ical Society, Chicago, Ill., Sept. 13-18, 1970.

CHAPTER 12

The Nucleic Acid Constituents Analyzed by High-Pressure Liquid Chromatography

Dennis R. Gere

A.   INTRODUCTION

The study of the biochemistry of nucleic acids is an exciting field of endeavor these days. The discovery of the structure of deoxyribonucleic acid (DNA) by Watson, Crick and Wilkins, followed by the elucidation of the mechanism of the genetic code by Holley, Nirenberg, and Rah Bandary has given great impetus to this interest. Advanced liquid chromatography instrumentation and techniques have important advantages over other analytical methods for the study of these complex and important compounds. High-performance liquid chromatography can provide the same degree of high resolution, high speed, and high sensitivity for the separation of nonvolatile compounds as gas chromatography does for volatile compounds.

This discussion will demonstrate applications of liquid chromatography instrumentation to the analytical separation of nucleotides, nucleosides, and N-bases. The analytical separations are carried out with mass quantities ranging from picograms to milligrams; the preparative separations, with mass quantities in the range of milligrams to grams. The liquid chromatography separation of higher-molecular-weight nucleic acids and oligonucleotides is generally done for preparative purposes and is the focus of a different area of technology.

B.   STRUCTURES OF NUCLEIC ACID CONSTITUENTS

In general, nucleic acids are discussed on five levels of structure: nucleic acid (DNA or RNA), oligonucleotides, nucleotides, nucleosides, and N-bases, with the highest (most complex) being the polymeric nucleic acid and the lowest (simplest) the individual N-base. Intermediate to these are the nucleosides (sugar glycosides of the N-bases), nucleotides (phosphate esters of the

417

nucleosides), and oligonucleotides (dimers and higher
polymers of nucleotides).

   Two main types of heterocyclic structures are found in
the N-bases. First are the pyrimidines; their parent
structure and the convention for numbering the positions
in the ring are as follows:

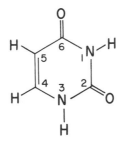

The prevalent pyrimidines found in nucleic acids are ura-
cil, thymine, and cytosine, with additional pyrimidine
derivatives being found in specialized nucleic acids,
such as the transfer RNA's. In addition to their importance in nucleic acids, pyrimidines and their nucleotides
play significant roles in carbohydrate, lipid, and vitamin metabolism.

   The second main type of N-base is the heterocyclic
purine structure,

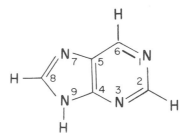

which contains a six-numbered pyrimidine ring fused to
the five-membered imidazole ring. Adenine and guanine
are the major purines of nucleic acid. Hypoxanthine and
xanthine are also found, although on a more limited
scale. These structures are as follows:

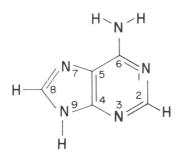

Adenine

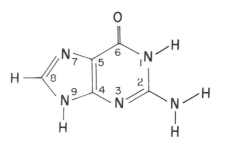

Guanine

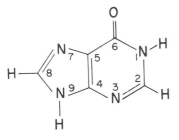

Hypoxanthine

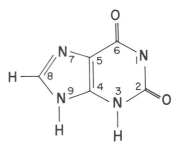

Xanthine

As with the pyrimidines, various methylated and other derivatives of purine are found in such compounds as transfer RNA's. In addition, purines have other important metabolic roles, and many plant purines, such as caffeine and theobromine, have important pharmacological functions.

The next higher substructures of nucleic acid are the nucleosides. These consist of an N-base that is chemically bonded in a β-glycosidic linkage with a pentose sugar. The nucleic acids are classified according to this sugar moiety, with RNA (left, below) containing the pentose ribose and DNA (right, below) the pentose deoxyribose:

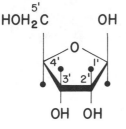

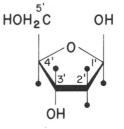

D-Ribose (β-D-ribofuranose)   D-2-Deoxyribose (β-D-2-
                               deoxyribofuranose)

As an example of a nucleoside, the structure of adenosine
is shown below:

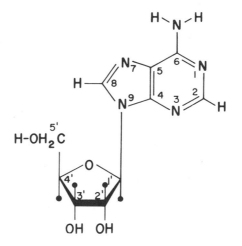

Adenosine (9-β-D-ribofuranosyladenine)

Nucleotides are the phosphate esters of the nucleosides. For the nucleotides containing deoxyribose, phos-
phorylation of the sugar is possible only at C-3' and
C-5', since C-1' and C-4' are involved in the furanose
ring and C-2' does not bear a hydroxy group. In RNA,
C-2', C-3', and C-5' isomers are found with only C-1'
and C-4' unavailable for esterification.

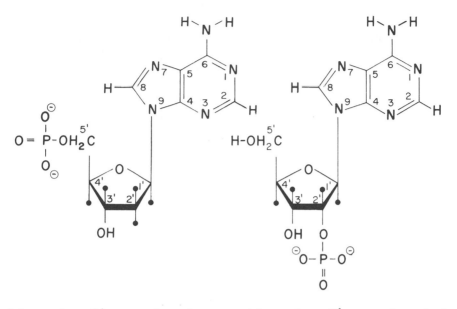

Adenosine 5'-monophosphate   Adenosine 2'-monophosphate

Which isomer a mixture contains depends on the mode of
hydrolysis, since different enzymes will give a mixture
of either 5' or 3' nucleotides from DNA and RNA, while
alkaline hydrolysis of RNA yields a mixture of 2' and 3'
nucleotides. In addition to their important role in
nucleic acid metabolism, the 5' monophosphates are found
free in muscle and other tissues. In addition, the 5'
monophosphates are further phosphorylated to di- and tri-
phosphates, many of which have important metabolic roles.
    The fundamental units of nucleic acid are the nucleo-
tides, starting with a dimer (dinucleotide) and culminat-
ing in a long-chain, high-molecular-weight polymer, the
nucleic acid. The repeating unit in these compounds is
the phosphodiester linkage between the 3' hydroxyl and
the 5' hydroxyl of neighboring sugar groups, that is,
ribose if the nucleic acid is RNA, and deoxyribose if
DNA. Since this sugar phosphate backbone is common to
all nucleic acids, the key to the specificity and function of the individual nucleic acids lies in the sequences
of the individual N-bases in the polymer chain. This was

demonstrated by the breaking of the genetic code in the
laboratories of Nirenberg, Holley, and Rah Bandary. For
further insight into the structure of the higher-molecular-weight nucleic acids, the reader is referred to the
many excellent texts covering this subject.

C. COLUMN AND INSTRUMENT TECHNOLOGY

Nucleotides, nucleosides, and the purine and the pyrimidine bases are, in general, nonvolatile, polar compounds which have reasonable solubility in aqueous solution. These compounds are ionizable in aqueous solution,
although the ionization is subject, in each case, to the
specific temperature, pH, and ionic environment of the
aqueous solution. All of these chemical characteristics
suggest that ion-exchange column chromatography is a very
functional and promising mode of separation. It is particularly important that the degree of ionization is subject to the three variables: temperature, pH, and ionic
environment. It will be shown how each of these parameters can be exploited to offer a virtually infinite potential for separating any given mixture of nucleic acid
constituents.

All of the analyses described herein were accomplished
on commercially available instruments, the Varian Aerograph model LCS 1000 and model 4100. The components and
features of the LCS 1000 have been described by Horvath
and others (1,2,3).

There are several constraints on the chromatographic
system which will be capable of the high-pressure, high-
speed, high-resolution, and high-sensitivity separation
of the subject compounds. The separating column must be
of relatively small diameter, and the ion-exchange separation materials should be capable of extremely fast
mass transfer kinetics, as indicated in Chapter 1. The
plumbing hardware must be of small internal volume. The
injection port, the detector, the column fittings, the
plumbing unions, and the transfer lines must all be of
such small volume that no mixing or remixing of separated compounds can take place to degrade the high-resolution separation of the column material. The ion-exchange

materials also must be mechanically stable at high pressures.

The resin used in the separations of the nucleotides, a strongly basic anion exchanger (and, in the case of the nucleosides and bases, a strongly acidic cation exchanger), was stable at pressures in the range of 3000-5000 psi. In addition, to incorporate fast mass transfer and rapid equilibration, the resins had a small particle diameter of 3-20 μ or, in the case of the pellicular type, a very thin resin depth. The fast mass transfer also increased the speed of regeneration at the end of a gradient elution separation. To ensure accurate identification of components, the resins used must exhibit the same retention volume for a given compound over a long period of time.

The detector should be capable of high sensitivity in order to sense a small mass of materials existing in biological samples. In addition, detectors must also have extremely good stability, both in short-term noise and long-term drift, as well as good reproducibility.

The ultraviolet (UV) detector is preferred for nucleic acid constituents. The detector utilized in the experiments described in this study was a single-wavelength UV photometer with a sensing wavelength of 254 nm. This detector has high sensitivity, is rugged, and is not very easily contaminated; hence the sensitivity will not vary appreciably over a long period of time. Most of the compounds encountered in the subject class of materials have molar absorptivities between 10,000 and 20,000. Compounds with a molar absorptivity of 10,000 can easily be detected at the 10^{-6} $\underline{M}$ level.

The mobile liquids used in these studies were aqueous solutions with pH ranging from 2 to 7. The buffer strength values ranged from 0.01 to 6.0 $\underline{M}$, and the compounds used included potassium dihydrogen phosphate, ammonium dihydrogen phosphate, ammonium formate, and sodium acetate.

The columns were made from stainless steel tubing. Most of them were 1/16 in.o.d. (1 mm i.d.) and 300 cm or 10 ft in length. Some of the separations were carried

out on 1/8-in.-o.d. (2.4-mm-i.d.) columns, 15 and 25 cm
in length.   In some instances, the first 6 in. of the
column was removable and was joined to the rest of the
column with a "zero-dead-volume" union. The purpose of
this detachable front end was to facilitate the regeneration of the first portion of the column, which often becomes contaminated with residues having a very low mobil-
ity in the separating system. This first portion of the
column is sometimes referred to as a precolumn, although
it is packed with the same material as the rest of the
column and is, in fact, an integral part of the separating column.

Liquid chromatography, like any other form of analytical chemistry, is a combination of art and science.
Therefore, it is not surprising that experimental conditions that give ideal results in one analysis sometimes
turn out to be less than perfect for another, similar
analysis. For example, changing from one production lot
of ion-exchange resin to another necessitates the changing of conditions to achieve optimum results. Horvath
(1),Uziel (4),Cohn (5), and Burtis (6-8) have described
several useful guidelines for determining which parameter
to adjust for optimum results in the separations described
here. As a general rule, the parameters given below
should be adjusted in the order in which they are listed.
Although adjusting a single parameter will often be sufficient, in some cases it may be necessary to change two
or more of them simultaneously or sequentially in order
to achieve optimum results.

1. pH of the Buffer Solution. Increasing or decreasing the pH by as little as 0.10 unit may be sufficient.

2. Column Temperature. Adjusting the temperature 5°
above or below the starting point will usually suffice to
indicate which change, if either, will be effective.

3. Ionic Strength of the Buffer Solution. Increasing
or decreasing the ionic strength by 0.1 $\underline{M}$ or less is
usually sufficient. Columns packed with pellicular ion-
exchange resins are characterized by a large ratio of
mobile phase to stationary phase; thus, the eluent
strength of the mobile phase can be much lower with a

pellicular ion-exchange column than with conventional
ion-exchange resins.

4. <u>Gradient Elution Program</u>. This variable is extremely useful, but its great potential also means that
one has to be quite judicious in the choice of the proper
conditions for separating a given mixture.

5. <u>Carrier Flow Rate</u>. To a certain point, decreasing
the carrier flow rate increases the resolution. Therefore, the flow rate is usually reduced by some 20% to
determine the effect. This change will also usually increase sensitivity. Obviously, reducing the flow rate
will cause a proportional increase in the analysis time.

6. <u>Type of Exchanging Ion</u>. This factor can affect a
drastic change in resolution, but the subject is too complex to cover here.

D. EXPERIMENTAL VARIATION OF PARAMETERS

Figures 12.1-12.5 illustrate the experimental variations of some of the parameters available in ion-exchange
column chromatography. These figures show the effect of
varying the particle diameter, the column length, the
major nature of the eluents, and the temperature for the
separation of nucleosides.

Figure 12.1 illustrates the separation of five nucleosides - uridine, inosine, guanosine, adenosine, and cy-
tidine - utilizing a strongly acidic cation-exchange
resin 7-10 μ in particle diameter. With this 25-cm column the average transit time through the column for a
nonreacting species is approximately 50 sec. The first
peak, uridine, elutes just beyond the holdup time of 50
sec, and the five nucleosides are completely resolved
and separated in 5 min.

The effect of reducing particle diameter on the separation of the nucleosides is shown in Figure 12.2. The
experimental conditions used were the same as those for
Figure 12.1 except that in this case the particle diameter ranged from 3 to 7 μ. Under these conditions, the
same five nucleosides are resolved in less than 4 min.
The price for this increased speed and efficiency is
higher pressure.

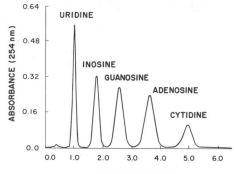

Figure 12.1. Fast Nucleoside Separation. Column
= 250 mm, 2.4 mm i.d.; packing = 7-10 μ cation-
exchange resin; carrier = 0.4M ammonium formate,
adjusted to pH = 4.75; temperature = 55°C; flow
rate = 35 cc/hr; inlet pressure = 4400 psi; carrier
velocity = 0.54 cm/sec.

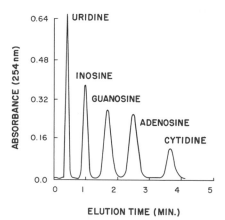

Figure 12.2. Reduced particle diameter. Conditions same as for Figure 12.1 except: cation-
exchange resin = 3-7 μ; carrier flow rate = 40
cc/hr; inlet pressure = 4800 psi.

The combination effect of changing column length, flow
rate, and temperature is illustrated in Figure 12.3.

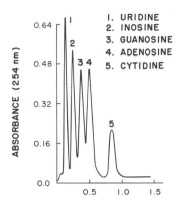

Figure 12.3. Higher temperature and shorter column. Conditions same as for Figure 12.2 except:
column = 15 cm long; temperature = 85°C; flow rate
= 60 cc/hr; inlet pressure = 4500 psi; carrier linear velocity 0.92 cc/min.

This chromatogram shows the application of the same 3-7
μ resin as above, but two conditions have been changed.
First, the column was shortened from 25 to 15 cm. This
allowed, at the same inlet pressure, a higher volume flow
rate. In addition, the temperature of the column was increased from 55 to 85°C. The transit time through the
15-cm column for nonretained species is now in the order
of 16 sec. Under these conditions the mixture of five
nucleosides is separated in less than 1 min. Increasing
the temperature has improved mass transfer, but the loss
in resolution for the first four peaks is due to the reduction in column efficiency caused by operating the
shorter column at a much higher carrier velocity.
    Resolution often may be improved by altering the pH,
as shown in Figure 12.4. In this separation the column
and the temperature were the same as for Figure 12.3,
but the pH of the 0.4M ammonium formate was lowered from
4.50 to 4.25. Resolution is significantly improved, with
the total analysis time increased from less than 1 min

to $2\frac{1}{2}$ min for the five nucleosides because of increased
retention.

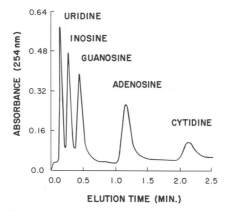

Figure 12.4. Lower pH. Conditions same as for
Figure 12.3 except: carrier = 0.4M ammonium formate adjusted to pH 4.25; flow rate = 55 cc/hr;
inlet pressure = 4500 psi.

Figure 12.5 shows another variation in the technique
to increase separation speed. In this case, a column of
larger-diameter resin particles (10-14 μ) was used. This
larger mesh size allowed a greater volume flow rate at
the same nominal inlet pressure. With these conditions,
four nucleosides are separated in 1 min and 15 sec. The
fifth nucleoside was not present in this mixture but
would appear, if it were, in the vacancy between uridine
and guanosine.

E.   ANALYSIS OF NUCLEOTIDES

The analysis of ribonucleotides and deoxyribonucleo-
tides can be carried out on a strongly basic anion-exchange resin. Since these nucleotides are related in
structure but vary greatly in ionization, a gradient elution technique is advisable for high resolution.

Figure 12.6 shows the separation of a standard mixture
of 2' and 3' nucleotides. Note that the order of elution
is, first, cytosine monophosphate, followed by uridine

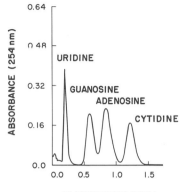

Figure 12.5. Larger particle diameter. Column = 250 mm, 2.4 mm i.d.; packing = 10-14 μ cation-exchange resin; carrier = 0.4M ammonium formate adjusted to pH 4.5; temperature = 85°C; flow rate = 85 cc/hr; inlet pressure = 4900 psi.

monophosphate, adenosine monophosphate, and guanosine monophosphate. In each case, the 2' isomer elutes before the 3' isomer. The total analysis time is less than 70 min, each peak representing less than a microgram of each component. Thus, with this system one can achieve high speed, high resolution, and high sensitivity. The column used for this separation was packed with pellicular anion-exchange resin. The volume flow rate of the mobile solvent was 12 cc/hr, and under these conditions an inlet pressure of 500 psi was generated. A gradient elution was utilized, and the starting eluent was 0.01M potassium dihydrogen phosphate adjusted to a pH of 3.35. To establish the gradient, 1.0M potassium dihydrogen phosphate adjusted to a pH of 4.3 was mixed into the starting solution at a volume flow rate of 6 cc/hr. The initial starting volume of the low-concentrate phosphate buffer was 50 ml, and the gradient onset was delayed for 7.5 min after the beginning of the analysis (the injection point). At the end of the separation a 35-min regeneration cycle was carried out, using the starting eluent.

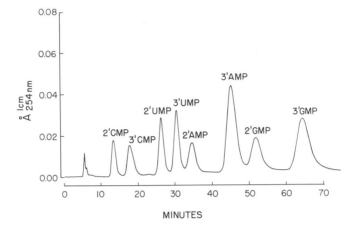

Figure 12.6. 2' and 3' nucleotides. Column = 300
cm, 1 mm i.d., pellicular anion-exchange resin;
gradient elution carrier, starting with 0.01$\underline{M}$ KH$_2$PO$_4$
adjusted to pH 3.35 at 12 cc/hr, modified with 1.0$\underline{M}$
KH$_2$PO$_4$ adjusted to pH 4.3 at 6 cc/hr; temperature
= 80°C; detector = 0.008 absorbance, full-scale.

Separation of the 5' nucleotides on an anion-exchange
resin column is illustrated in Figure 12.7. With half
as many compounds in this mixture as in the 2',3' nucleo-
tide mixture in Figure 12.6, less total resolution is
required and the analysis time can be significantly reduced to 12 min. A faster column flow rate and a much
steeper gradient were utilized to accomplish the 12-min
analysis. However, a 20-min regeneration cycle was necessary between each gradient run. Hence, the total recycle
time is 32 min per analysis. It was observed that each
deoxyribonucleotide has a slightly greater relative retention than its ribonucleotide analog. It is also interesting to note that in all cases the 5' isomer elutes
before the 2' isomer, which in turn elutes before the 3'
isomer. The starting eluent and the gradient eluent were
of the same composition as in Figure 12.6, but the initial

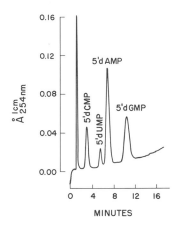

<u>Figure 12.7</u>. 5' nucleotides. Conditions same as
in Figure 12.6 except: temperature = 75°C; starting inlet pressure = 2800 psi; gradient as described in text; UV detector = 0.16 absorbance,
full-scale.

volume was only 20 cc. To this initial volume of low-
concentration eluent, a gradient mixture was added at
the rate of 24 cc/hr. There was no gradient delay in
this analysis. Each peak represents 1 μg of each component in the chromatogram.

Another type of separation which is often necessary
in the analysis of nucleic acid constituents is shown in
Figure 12.8. Adenosine monophosphate, adenosine diphos-
phate, and adenosine triphosphate are separated with a
pellicular anion-exchange packing. In this example, a
linear elution separation (no gradient) was carried out
in less than 4 min. Not only was this a fast separation
but also, because a nongradient elution was used, the regeneration cycle was unnecessary. Thus, the total recycle
time for each separation is 4 min.

F. ANALYSIS OF NUCLEOSIDES AND N-BASES

In many respects, the nucleosides represent an easier
family of compounds to resolve than the nucleotides, as

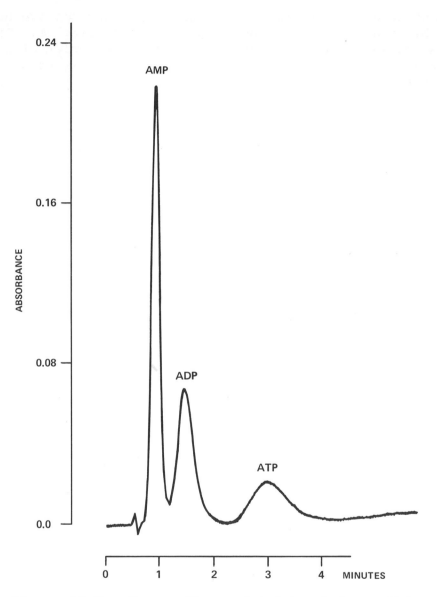

<u>Figure 12.8</u>. Mono-, di-, and triphosphates. Col-
umn as in Figure 12.6; carrier = 0.75$\underline{M}$ KH$_2$PO$_4$ adjusted to pH 4.2; temperature = 80°C; flow rate =
55 cc/hr; inlet pressure = 2800 psi; UV detector
= 0.16 absorbance, full-scale; sample = 10 μg of
each component.

manifested by the fact that nongradient elutions are possible even for pairs that are difficult to resolve. Time can be saved if a regeneration cycle (a necessity with gradient elution) can be avoided. The nucleosides or ribosides are similar in their chemical characteristics to the purines and pyrimidine bases because the very weakly acidic carbohydrate substituents have only a secondary effect on the ionization of the basic purine and pyrimidine moieties. Because of the absence of the phosphate groups found on the nucleotides, one is not faced with a separation of a variety of isomers, as in the case of the 2' and 3' nucleotides, the 5' nucleotides, and the cyclic nucleotides. For these reasons, the conditions needed for use with an ion-exchange column to separate the nucleosides are less drastic.

A typical separation of four nucleosides using a pellicular ion-exchange resin is shown in Figure 12.9. In this illustration, a strongly acidic cation-exchange resin was employed.

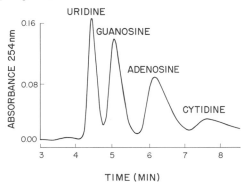

Figure 12.9. Nucleosides. Column = 300 cm, 1 mm i.d., pellicular strong cation exchanger; carrier = $0.02\underline{M}$ $NH_4H_2PO_4$ adjusted to pH 5.0; carrier flow = 14.3 cc/hr; inlet pressure = 725 psi; temperature = 40°C.

Figure 12.10 illustrates the separation of some deoxynucleosides. This particular mixture represented the materials present in the cells of a rare species of crab. The mole ratio of (deoxyadenosine plus thymidine) to (deoxyguanosine plus deoxycytidine) was 30 to 1. This

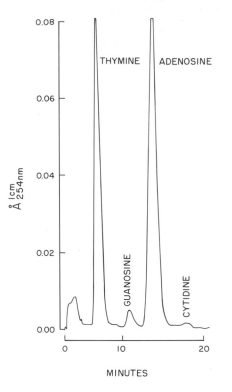

Figure 12.10. Deoxynucleosides. Column = 250 mm, 2.4 mm i.d., Bio-Rad A-7 strong cation-exchange resin, 7-10 μ; carrier = 0.4M ammonium formate adjusted to pH 4.55; temperature = 55°C; flow rate = 10 cc/hr; inlet pressure = 1000 psi.

situation placed great stress on the chromatographic system because the thymidine peak elutes first, followed by deoxyguanosine, deoxyadenosine, and finally deoxycytidine. The size of these peaks (disregarding molar response factors and band-broadening effects) was 30 to 1 to 30 to 1. With the sample consisting of alternating large and small peaks, a conventional (totally porous) ion-exchange resin was selected (as opposed to a superficially porous or pellicular ion-exchange resin) to provide maximum capacity and resolution. The resin used was Bio-Rad® A-7

strongly acidic cation-exchange material with a very narrow particle distribution (95% of all particles fell
within the range of 7-10 μ). The operating conditions
were chosen in accordance with the guidelines discussed
in Sections C and D. It is possible that there are other
sets of conditions which would provide equivalent results.

The purine and pyrimidine bases of the nucleic acids
are shown separated in Figure 12.11. This very fast separation of the N-bases was achieved with a pellicular
ion-exchange resin. The inlet pressure for this 5-min
separation was 2800 psi.

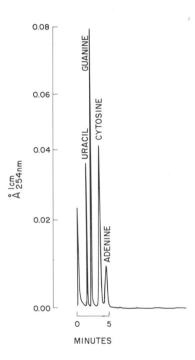

Figure 12.11. Purine and pyrimidine base separation. Column = 250 cm, 1 mm i.d., pellicular cation-exchange packing; carrier = 0.025M NH4H2PO4;
temperature = 70°C; carrier flow rate = 36 cc/hr.

G. URINE ANALYSIS

The chromatogram in Figure 12.12 illustrates the application of high-performance liquid chromatography to the analysis of UV-absorbing constituents (such as nucleic acid constituents) of human urine. For this separation, a strongly basic anion-exchange resin (particle diameter 12-15 µ) was used with gradient elution to accomplish the desired analysis. The beginning eluent was 0.015M sodium acetate buffer adjusted to a pH of 4.4. To this was fed a gradient mixture of sodium acetate at a concentration of 6.0M and a pH of 4.4. A linear gradient was generated, starting 30 min after the point of injection and continuing in a linear mode for the first 24 hr of analysis. From 24 to 30 hr, the analysis was a nongradient eluent of the final 6M concentration. The volume flow rate of the mobile liquid was 8 ml/hr with an inlet pressure ranging from 1000 to 1600 psi. The temperature of the column was 25°C for the first 4 hr and then was increased to 60°C for the remaining 26 hr. Two hundred microliters of untreated human urine was injected into the column through a sample loop. In the 30-hr analysis time, some 90 peaks were obtained; only a few of them have been identified to date. Uric acid elutes at 5.3 hr, for example, and hippuric acid at 13.4 hr. The identity of these compounds was determined by comparing their elution times with those of known compounds.

H. SUMMARY

Since Cohn first separated the N-bases in 1949, liquid chromatography has been an important tool for the separation and quantitation of nucleic acids and their components. For the separation of nucleotides, anion-exchange chromatography [Cohn (4,9), Volkin et al. (10), and Caldwell (11)] has been the most popular method, but cation-exchange chromatography has also been used [Katz and Comb (12), Manley and Manley (13), Blattner and Erickson (14), and Junowics and Spencer (15)]. Anion-exchange chromatography has been employed for the separation of N-bases and nucleosides [Green et al. (16), and

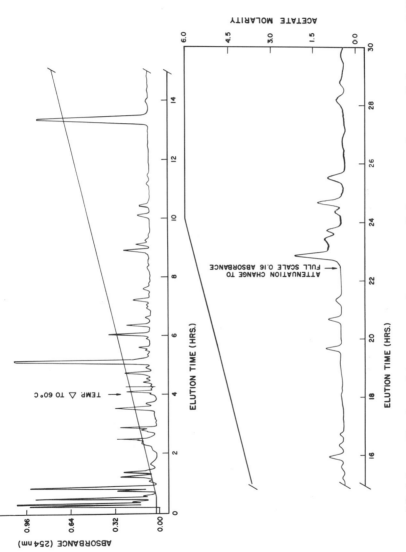

Figure 12.12. UV-absorbing constituents of human urine. Column = 100 cm, 2.4 mm i.d., strong anion-exchange resin, 12-15 μ; gradient elution and other conditions as described in text.

437

Anderson et al. (17)] but cation-exchange chromatography
has proved to be more effective [Crampton et al. (18),
Busch (19), Uziel et al. (4), Junowics and Spencer (15),
Horvath and Lipsky (2), and Burtis et al. (6)]. With the
exception of the methods of Uziel et al., Horvath et al.,
and Burtis et al., the separations cited above were time
consuming, requiring from 1 to several hours.

The chromatograms in this chapter demonstrate that the
time required for separation of the components of nucleic
acids and body fluids can be greatly decreased by using
high-pressure liquid chromatography with high-efficiency
chromatographic columns and sensitive UV detection.

References

1. C. Horvath, B. Preiss, and R. R. Lipsky, Anal. Chem.,
    39, 1422 (1967).
2. C. Horvath and S. R. Lipsky, Anal. Chem., 41, 1227
    (1969).
3. Varian Aerograph LCS 1000 Instruction Manual.
4. M. Uziel, C. K. Koh, and W. E. Cohn, Anal. Biochem.,
    25, 77-98 (1968).
5. W. E. Cohn, Science, 109, 377 (1949).
6. C. A. Burtis, D. R. Gere, J. M. Gill, and F. M.
    MacDonald, Chromatographia, in press.
7. C. A. Burtis, J. Chromatog., paper submitted for
    publication.
8. C. A. Burtis, Fed. Proc., 29, 726 Abs. (1970).
9. W. E. Cohn, in "The Nucleic Acids: Chemistry and
    Biology," Vol. 1, F. Chargaff and J. N. Davidson,
    eds., Academic, New York, 1955, Chapter 10.
10. E. Volkin, J. X. Kymn, and W. E. Cohn, J. Am. Chem.
    Soc., 73, 1533 (1951).
11. I. C. Caldwell, J. Chromatog., 44, 331 (1969).
12. S. Katz and D. G. Comb, J. Biol. Chem., 238, 3065
    (1963).
13. F. Manley and G. J. Manley, J. Biol. Chem., 235
    (1960).
14. F. R. Blattner and H. P. Erickson, Anal. Biochem.,
    18, 220 (1967).

15. E. Junowics and J. H. Spencer, J. Chromatog., _37_, 518 (1968).
16. J. C. Green, C. E. Nunley, and N. G. Anderson, Natl. Cancer Inst. Monogr. _21_, 431 (1966).
17. N. G. Anderson, J. G. Green, M. L. Barber, and F. C. Ladd, Sr., Anal. Biochem., _6_, 153 (1963).
18. C. F. Crampton, F. R. Frankel, A. M. Benson, and A. Wade, Anal. Biochem., _1_, 249 (1960).
19. E. W. Busch, J. Chromatog., _37_, 518 (1968).

List of Symbols

A	= eddy diffusion term in plate height equation
A_R	= proportionality constant, called the response factor
c	= concentration of solute in the mobile phase
C	= mass transfer term in plate height equation
C'	= constant, independent of particle diameter
C_S	= stationary phase mass transfer term
C_M	= mobile phase diffusion mass transfer term
C_M^*	= stagnant mobile phase diffusion mass transfer term
D	= detector output
D_M	= diffusion coefficient of solute in mobile phase
D_S	= diffusion coefficient of solute in stationary phase
d	= diffusion path length (equated with effective film thickness of stationary liquid phase)
d_c	= internal diameter of the column
d_p	= particle diameter of the packing
f	= total porosity in the column
F	= mobile phase flow rate
$\Delta(\Delta G^\circ)$	= differences in standard free energies of distribution of two solutes
h	= reduced plate height (H/d_p)
H	= height equivalent to a theoretical plate
H_{eff}	= height equivalent to an effective plate
H_{min}	= minimum H, occurring at the optimum velocity

ΔH^m	=	heat of mixing two liquids
K	=	distribution coefficient (concentration ratio)
K^o	=	specific permeability of a column
K'	=	permeability constant
K_o	=	bulk coefficient for gel permeation chromatography: volume fraction of the stationary phase available to solvent molecules/the volume of the stationary phase
k'	=	capacity factor (amount or partition ratio), KV_S/V_M
k_d	=	rate constant for desorption
L	=	column length
M_w	=	molecular weight
N	=	number of theoretical plates
N_{eff}	=	number of effective plates
N_{eff}/t	=	effective plates per second
q	=	configurational correction factor for differing pore shapes
PLB	=	porous layer beads
P_m	=	polarity of the mobile phase
P_s	=	polarity of the stationary phase
ΔP	=	column pressure drop
ΔP_{lim}	=	apparatus limiting pressure drop
R	=	gas constant
R_s	=	resolution
T	=	temperature, °A
t	=	time of analysis

t_R	= retention time of a peak measured from the start
t_o	= retention time of a non-sorbed species (hold-up time)
t'_R	= $t_R - t_o$ = reduced retention time
t_a	= mean sorption time
t_d	= mean desorption time
v	= mobile phase velocity
v_{min}	= v at the beginning of the H/v versus v curve minimum
v_{opt}	= optimum efficiency velocity at which H = H_{min}
v'_{opt}	= optimum velocity for fast analysis
V_R	= retention or elution volume
V_R^o	= specific retention volume
V_M	= column mobile-phase or interstitial volume
V_S	= column stationary-phase volume, or volume of solvent in the pores of an exclusion packing
V_N	= net retention volume
V_T	= total elution volume V_R when $K_o = 1$
V_{fs}	= volume of stationary phase available to solvent
V_i	= molal volume of liquid i
V_j	= molal volume of liquid j
w	= peak width
X_i	= mole fraction of liquid i
X_j	= mole fraction of liquid j
X_m	= equilibrium distribution of sample component X in mobile phase

X_s = equilibrium distribution of sample component X in stationary phase

α = relative retention, k_2'/k_1'

ν = tortuosity correction factor for mobile phase diffusion

ν_n = number of cycles in recycling chromatography

δF = fractional film thickness of porous layer of PLB

δ = Hildebrand solubility parameter

$\delta_a, \delta_d, \delta_h, \delta_x, \delta_s, \delta_m$ = individual terms of the Hildebrand solubility parameter

ϵ = interparticle porosity

$\epsilon°$ = solvent strength parameter

η = mobile phase viscosity

λ = packing irregularity correction factor in eddy diffusion term

σ = standard deviation of Gaussian band

σ_t = time-based standard deviation

σ_L = length-based standard deviation

$\Theta_{.1}$ = adsorbent linear capacity

σ_T = total standard deviation

σ_{col} = standard deviation of band broadened in the column

σ_{ex-col} = standard deviation of band broadened by extra-column effects

ϕ_i = volume fraction of liquid i

ϕ_j = volume fraction of liquid j

Φ = fraction of mobile phase occupying intraparticle space

ϕ = peak capacity

ω   = dimensionless constant, weighting factor for
              mobile-phase diffusional band broadening

$\searrow$ = reduced velocity

μ   = micrometer

μg   = microgram

μl   = microliter

INDEX